Poison Powder

environmental history and the american south

SERIES EDITORS

James C. Giesen, Mississippi State University

Erin Stewart Mauldin, University of South Florida

ADVISORY BOARD

Judith Carney, University of California–Los Angeles

S. Max Edelson, University of Virginia

Robbie Ethridge, University of Mississippi

Ari Kelman, University of California–Davis

Shepard Krech III, Brown University

Megan Kate Nelson, www.historista.com

Tim Silver, Appalachian State University

Mart Stewart, Western Washington University

Paul S. Sutter, founding editor, University of Colorado Boulder

Poison Powder

THE KEPONE DISASTER IN
VIRGINIA AND ITS LEGACY

Gregory S. Wilson

The University of Georgia Press
Athens

© 2023 by the University of Georgia Press
Athens, Georgia 30602
www.ugapress.org
All rights reserved
Set in 10.5/13.5 Garamond Premier Pro
by Kaelin Chappell Broaddus

Most University of Georgia Press titles are
available from popular ebook vendors.

Printed digitally

Library of Congress Cataloging-in-Publication Data

Names: Wilson, Gregory S., author.
Title: Poison powder : the Kepone pesticide disaster in Virginia and its
 legacy / Gregory S. Wilson.
Description: Athens, Georgia : The University of Georgia Press, [2023] |
 Series: Environmental history and the American South | Includes
 bibliographical references and index.
Identifiers: LCCN 2022037313 | ISBN 9780820363486 (paperback) |
 ISBN 9780820363479 (hardback) | ISBN 9780820363493 (epub) |
 ISBN 9780820364032 (pdf)
Subjects: LCSH: Chlordecone—Toxicology—Virginia—Hopewell—
 History. | Chlordecone—Toxicology—History. | Chlordecone—
 Environmental aspects—Virginia—Hopewell—History. |
 Chlordecone—Environmental aspects—West Indies, French—History.
 | Allied Chemical Corporation—History. | Pesticides industry—
 Environmental aspects—Virginia—Hopewell—History. | Pesticides—
 Government policy—United States—History. | Environmental
 health—United States—History. | Environmental justice—United
 States—History.
Classification: LCC RA1242.C43 W55 2023 | DDC 363.738/409755586—dc23/
 eng/20220805
LC record available at https://lccn.loc.gov/2022037313

CONTENTS

Foreword vii

Acknowledgments ix

INTRODUCTION 1

CHAPTER 1. The James River before Kepone 12

CHAPTER 2. Hopewell, Allied, and the Beetle Battles 24

CHAPTER 3. The Kepone Shakes and a Poisoned River 45

CHAPTER 4. Biocitizenship and Accountability 68

CHAPTER 5. A Crime against Every Citizen 94

CHAPTER 6. Kepone and the Environmental Management State 127

CHAPTER 7. The Present and Future of Kepone 157

EPILOGUE 181

Notes 189

Index 223

FOREWORD

Poisons occupy a somewhat predictable place in both academic and popular understandings of the past. Many common narratives portray insecticides and herbicides as disasters waiting to happen, and there are dozens of toxic histories that bear this out. Calcium arsenate killed boll weevils but also poisoned sharecroppers. DDT helped to control malaria but killed fish and the birds that ate them. Malathion wiped out mosquitoes but sickened some people spraying it. Again and again, people have been caught unaware when the dangerous chemicals they use start to have unintended and deleterious effects on the flora and fauna around them. Sometimes it comes as a surprise when a chemical that has been engineered to poison things starts poisoning things. At other times, manufacturers have hidden important details. Governments have left the offending chemicals unregulated or failed to enforce their own rules. Public health and environmental protection groups have been slow to react. This narrative structure of toxic disasters has become a trope of environmental research.

The history in this book puts much of what we know about poison production, use, and disaster to the test. What follows is a story about the manufacture and application of a powdery pesticide known generically as chlordecone, produced in Hopewell, Virginia, under the brand name Kepone. But there is very little in this narrative that goes as one might expect. There was no single major accident, spillage, or mass poisoning. There was no revolutionary moment of discovery about the chemical's danger to humans or wildlife. This book is not about a long-silent killer oozing up from below to harm people or poison animals, or powerful chemical companies strong-arming government regulators. Kepone's history has elements of all of these, to be sure, but it is different from the common telling both in its narrative arc and its lessons for readers.

As Gregory Wilson demonstrates, understanding Kepone's effects means untangling a complex web of factors surrounding all aspects of the pesticide's manufacture and use. Using a creative combination of oral histories, legal findings, and regulatory case files, Wilson shows how environmental damage and human health impacts occurred seemingly everywhere the chemical touched, from its production plant along the James River to its transportation to fields where it was dusted on potato and banana plants across the United States, Europe, and in the Caribbean. Wilson employs the term "residues" to show how the toxin's impacts on bodies, waterways, plants, and animals were both immediate and very long-term. The Kepone disaster left behind legal and regulatory residues as well. It left its mark in the psychology of those affected by it. Those impacts, Wilson shows, were determined not only by proximity to the poison but also to some extent by distance from regulators, health care expertise, and functioning state supervisors. Kepone's damage and lasting effects were not the ones we expect, however. As Wilson shows, many of the people poisoned by the chemical seem to have gotten better. Readers may be surprised at moments in the book when the state arrives, quickly and effectively protecting workers and the environment. However, this is still a disaster story, and Wilson demonstrates how race, geography, and politics were determinants of toxicity and shaped damage, protection, and recovery. For every industrial worker whose involuntary spasms—"the Kepone shakes"—subside, there are farmers thousands of miles away who are not getting better, living among lands embedded with Kepone that will last generations.

Poison Powder sits at an important nexus of themes within the Environmental History and the American South series and southern environmental history generally. It speaks to the importance of waterways, rural manufacturing, and agriculture. Indeed, the book reminds southern environmental historians how central agricultural products have been to the region's development, and not merely in the farm field. Kepone's history taps into a far-reaching economic system that incorporates chemical producers as well as consumers. Farmers' needs to beat back insects spurred the development of the chemical companies of coastal Virginia that made the pesticide, and Kepone's history is a reminder of how central themes of labor, race, and state regulation are to all of these stories.

James C. Giesen and Erin Stewart Mauldin, editors
Environmental History and the American South

ACKNOWLEDGMENTS

I suppose this book began in 1976 when, as a child, I moved to Newport News, Virginia. For some reason, one of the many memories I have of that moment was hearing about a poison in the James River called Kepone. Essentially a lifetime later, as I started teaching environmental history, I began to recall my old hometown and the issue of Kepone in the water. What happened to it, I wondered? Well, one thing led to another, and the result is this book. Along the way, many individuals and institutions provided their support, encouragement, and cooperation.

The University of Akron awarded me a Faculty Research Grant in 2015 that made possible the early research for the project and then professional development leave in 2016–17. I thank my colleagues and the provost's office for showing me support. In 2016 I received a fellowship from the Virginia Foundation for the Humanities (VFH) as well as an Andrew W. Mellon Research Fellowship from the Virginia Museum of History & Culture (VMHC). Both enabled me to conduct the bulk of the research for the book, and without their support I could not have completed this project. At the VFH I am indebted to founder and former president Rob Vaughan as well as fellowship director Jeanne Siler. Jeanne helped in so many ways, including ensuring that I was safely ensconced in Richmond, Virginia, for my research in the fall of 2016. At the VMHC I thank John McClure for orienting me to the museum's fellowship program and the collections.

In Richmond the staff at the VMHC were extremely helpful in locating records in their holdings related to Kepone, and I am grateful for their help. The VFH fellowship provided an office at the Library of Virginia (LVA). While there I had the distinct pleasure of working closely with three fantastic historians: John Deal, Mari Julienne, and Brent Tarter. John served as my guide to the LVA and liaison with the VFH. Mari Julienne and Brent Tarter took time away from their own projects to help me navigate both the archives and Virginia history. All of them listened with patience and offered sound advice as I worked through my ideas on Kepone. They also read early portions of the book. While at the LVA, I received excellent assistance from archivists and librarians across the facility. I owe special thanks to Roger Christman, who uncovered several boxes of records from the Virginia's State Records Center and secured their transfer to the LVA. Ginny Dunn and Minor Weisiger helped me locate several collections and other records related to Kepone. The LVA also hosted my first public talk on Kepone that fall. I also thank fellow VFH award winner Kate Jones and her husband Casey, who were also in Richmond. Their friendship and kindness only made the stay better.

Further support came from Joseph Maroon, executive director of the Virginia Environmental Endowment (VEE). Joe shared his own Kepone story, and I worked with him in writing an editorial in response to federal efforts to end legal settlements that included supplemental environmental projects in enforcement. This type of settlement from Kepone created the VEE in 1977. Joe also worked with Graham Dozier at the VMHC to provide me the opportunity deliver a talk in 2017 as part of the Banner Lecture series and the VEE's fortieth anniversary.

I also thank the staff at Virginia's Department of Environmental Quality (DEQ) who helped me uncover and understand several records on Kepone that remained in their office in Richmond. While there, I was able to get the assistance of (and interview) former DEQ director David Paylor as well as Dennis Treacy, who, among his many roles, also served as DEQ director.

While working on this book, I had the pleasure of participating as a historian in a documentary on chlordecone for French television by filmmaker Bernard Crutzen, *Pour quelques bananes de plus: Le scandale du chlordécone*. Bernard and I shared our knowledge of chlordecone and the Kepone crisis, and I am forever grateful for meeting him and for the opportunity to understand more deeply the French and Caribbean aspects to this story. Through Bernard I was able to meet Richard Foster, whose writing on Kepone in *Richmond Magazine* helped me so much.

In Virginia, I also made extensive use of the Herbert H. Bateman State Senate Papers at the Special Collections Research Center at the College of William & Mary in Williamsburg, Virginia. I thank the staff there for helping me navigate those records. In Charlottesville, I made excellent use of the papers of C. Brian Kelly, who was a reporter and writer covering the Kepone disaster. I thank the staff at the University of Virginia's Special Collections Library for their help.

Part of my archival journey took me to the Philadelphia branch of the National Archives and Records Administration and the main office in Maryland. I thank the staff at both locations for their help. Many of the federal legal records generated from the Kepone disaster are gone, but fortunately the NARA branch in Philadelphia has retained some, which were essential.

Getting images for use in books can be daunting. Thanks to Tony Dudek from the *Richmond Times-Dispatch* and Michelle Gullett from the *Daily Press* (Newport News, Va.). Mark Merhige was particularly helpful in securing permission for a photograph of his father and sharing his memories of Judge Merhige.

Speaking of interviews: oral history forms a critical foundation of the book. I owe a special debt to two of Hopewell's former city managers, Mark Haley and Clint Strong, who sat down for interviews and showed me around the city. Jeanie LeNoir Langford shared both her Kepone story and her deep knowledge of Hopewell history. I am grateful for the willingness of these and all the narrators to open up about their experiences and memories. They offered a variety of viewpoints and recollections that paper records could never capture. These memories—some of them visceral—remind us of the human dimension at the heart of the Kepone story.

Being back in Virginia allowed me to reconnect with friends and family and explore other aspects of Virginia history and culture. Their support during the research and writing has been essential and rewarding. A big thanks to my sister Karen, her husband David, and my nephew Skyler Hutcheson. My fellow historian and friend Bil Kerrigan took time away from his own research to spend an afternoon exploring the Fredericksburg and Spotsylvania battlefields. Mike Fraughnaugh and his wife Lany Prousalis were consummate hosts in Richmond. Joining us as often as he could was Andy Beasley. I am ever grateful for their generosity and support and our far-ranging conversations and explorations of the region's music, food, and culture. I also must mention Bill Conner and Tom Walz, friends with whom I was able to reconnect.

At the University of Georgia Press, Jim Giesen and Mick Gusinde-Duffy were supportive from the beginning. They heard me deliver a conference presentation about Kepone and urged me to publish my research. They helped me navigate the process, and their sharp analysis helped improve the manuscript along the way. Beth Snead and Lea Johnson steered me through production. Tobiah Waldron did the indexing.

Finally, I dedicate the book to my wife Laura Hilton and my daughter, Kate Wilson. Busy with her own work as a historian, Laura took on the burden of running the household while I was in Richmond and during the process of writing offered her own wise counsel. Without her support, this book would not be possible. My daughter Kate was in elementary school when I started the research for this book and is in high school as this book goes to print. She has always reminded me that there are more important things than work. Yet for whom do we write history? Not for ourselves alone but also for the next generation.

Poison Powder

INTRODUCTION

It was muggy and still dark at 4:00 a.m., July 1975. Lieutenant R. L. Anderson of the Hopewell (Virginia) Police Department was in his squad car on a routine patrol. Hopewell was (and is) an industrial city of some twenty-three thousand residents, situated at the confluence of the Appomattox and James Rivers some twenty-five miles southeast of Richmond, the state capital. The officer likely drove southeast through the town on Randolph Road, the main thoroughfare, and State Route 10. Anderson no doubt understood why the sign on the side of this road welcomed drivers to the "Chemical Capital of the South." On his left, he passed by Life Science Products (LSP), a small chemical operation on a half-acre site in a converted gas station. Just beyond Life Science lay Allied Chemical, one of the world's largest chemical firms, whose sprawling plant along the James River was a warren of pipes and buildings.

As Lieutenant Anderson drove past the main Allied factory, he looked to his right and passed by what he and other locals called the "pebble plant." It had been Allied's Pebbled Ammonium Nitrate Plant where the company made nitrogen fertilizer, and it sat just across Randolph Road from Allied's main facility. On land that was part of the pebble plant, Anderson noticed two men standing somewhat off the road near the back of their pickup truck. It is not known if he talked with the men, nor if he knew them. But he observed that the pickup truck had a large tank filled with a liquid that the men were dumping into a large pit. These two men probably did what Life Science worker Frank Arrigo remembers doing on a similar trip. As his coworker backed the truck up to the pit, "not a lined pit, just a hole in the ground," they got out, "opened that valve up and just sat around talking until that whole thing drained." Arrigo remembers, "That stuff was smoking," and he says the plume was visible from Route 10.[1] Unsure whether what he saw was authorized, the officer filed a report with city officials after finishing his patrol. Nothing was done.

The men were dumping liquid waste from Life Science that contained Kepone along with other chemicals used to make it, likely hexachlorocyclopentadiene (HCP) and sulfur trioxide (SO_3). Kepone is the brand name for the powdered pes-

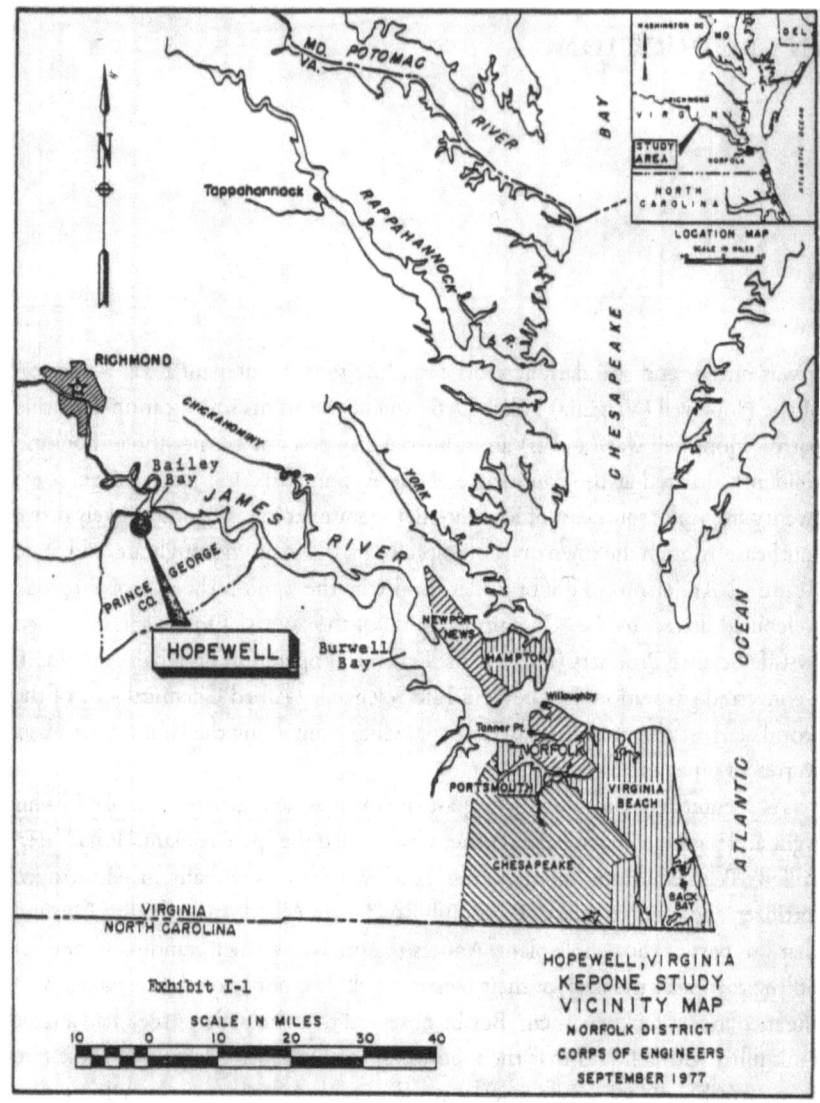

Map of Hopewell, Virginia, vicinity. This map, which shows Hopewell's location along the James River, was included in the EPA's 1978 report on possible mitigation efforts to remove Kepone from the river.

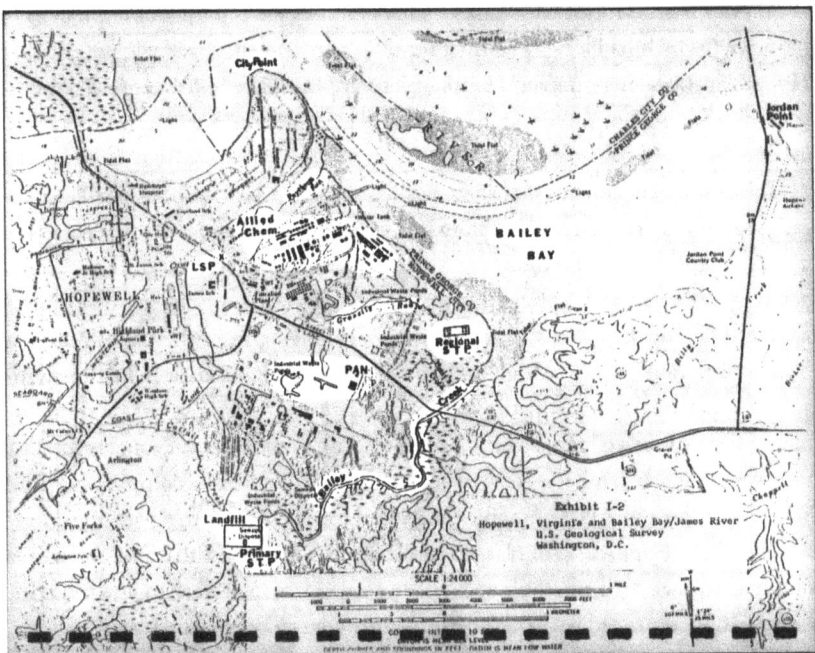

Map of Hopewell, Virginia, and Bailey Bay (James River). This map was also included in the 1978 mitigation study. It shows the location of several important facilities in the Hopewell disaster. The Allied Chemical facility is along Bailey Bay. The Life Science Products (LSP) building is just a little southwest from Allied, along Route 10 (labeled Randolph Street in this map). Along Route 10 southeast is the Pebbled Ammonium Nitrate facility (PAN), or "the pebble plant," once owned by Allied and where workers dumped Kepone waste. Nearby is the Regional Sewage Treatment Plant currently in operation, at the confluence of Bailey Creek and the James River. Farther south and situated along Bailey Creek and Cattail Creek is the defunct Primary Sewage Treatment Plant, whose digesters shut down from Kepone waste, and the landfill where workers dumped Kepone waste. After the disaster, remnants of Life Science were buried at the landfill and at the now closed Primary Plant.

ticide chlordecone, and Life Science held an exclusive contract with Allied Chemical to manufacture it. The production of Kepone involves first combining HCP and sulfur trioxide in a reactor with the catalyst antimony pentachloride. This liquid is transferred to quench tanks, where water, base, and acid are added to produce a gelatinous, cake-like precipitate of Kepone. This compound is then filtered, the soft Kepone cut, and the pieces dried into a fine white-to-tan powder that is then loaded into barrels for shipment.[2] As Arrigo's account suggests, what Anderson saw that predawn morning was not the first trip that pickup truck fixed with a 250-gallon tank had made to the pit. Over time, workers dumped perhaps as many as thirty or more tanks there.³ Indeed, such practices were not uncommon in 1975 Hopewell, which had long been the butt of jokes: "I smell, you smell, we all smell Hopewell." Local historian and Hopewell native Jeanie LeNoir Langford remembers those jokes. Yes, she says today, it may have smelled, "but that's the smell of money."⁴

These were valuable jobs, particularly for those workers without a college degree who populated the factory floors of plants like Allied or Life Science. This was especially true in the mid-1970s when the economy sputtered. From October 1973 to March 1974, prices for oil and gas skyrocketed from the first oil crisis of the decade. Economist Paul Samuelson told Americans they were suffering from "stagflation," a mix of high unemployment and high inflation. National unemployment had increased from 4.9 percent in 1973 to 8.7 percent by the summer of 1975. Hopewell fared a little better, with unemployment at 7.9 percent. But that was higher than it had been before the recession. So when a company like Life Science was hiring, it was an opportunity and for some perhaps a way to gain experience before applying to a larger company like Allied. With some thirty-five hundred workers, Allied employed more people in Hopewell than any other industry. They combined with other large companies—Continental Can, Firestone, and Hercules Powder—to provide a total of seven thousand jobs, occupy eleven hundred acres, and produce $97 million in annual payroll.⁵ Not surprisingly, the workforce in these companies reflected the norms of the era. Managers and plant workers were white men, the difference being that upper-level managers possessed a college degree. Some African American men got jobs in these plants but usually as janitors or laborers. White women dominated the secretarial pool and were office managers.

In November 1973, as the Watergate hearings were underway in Washington, D.C., the State of Virginia granted a corporate charter for Life Science Products to two former Allied Chemical experts, Virgil A. Hundtofte and William P. Moore. Hundtofte would be LSP's plant manager, and Moore its president. Moore had recently retired from Allied as director of agricultural research, but the company retained him as a consultant. Like Moore, Hundtofte also had worked for Allied, retiring about a year earlier from his position as the plant manager for Allied's agricultural production in Hopewell. With a corporate charter now in hand, on November 30 they signed a tolling arrangement with Allied to produce Kepone, meaning that Allied would supply the "chemical constituents of Kepone" and that Allied would buy all the Kepone LSP could make.⁶ Allied held the patent for the pesticide and produced it in Hopewell from 1966 to 1974. Allied had made similar arrangements before to make Kepone, with Nease Chemical in State College, Pennsylvania, and Hooker Chemical in Niagara Falls, New York, in the 1960s.⁷

In producing Kepone, Life Science did more than dump toxic waste into a large pit. The company sent Kepone waste into the city's sewage system, disrupting the treatment plant, and workers also dumped bad batches of Kepone directly into the sewer drains near the plant and at the city landfill. Initially, regulators were unaware of Life Science's existence. When they first discovered the pollution, they allowed the company to continue dumping as they looked for a technical fix to the problem. As this went on, another crisis brewed inside the Life Science plant. Workers complained of aches, pains, nervousness, and tremors they called the "Kepone shakes," or quivers, as lax safety standards led to Kepone poisoning. When they were able to see doctors that would thoroughly examine them, the prognosis was dire. One of the Life Science workers, Tom Fitzgerald, remembered: "One of the concerns was that I was sterile, and it's like any other poison. The symptoms were the same essentially as long-time drinking or heroin addiction. Any kind of poison damages your liver, your pancreas, your reproductive organs, your nervous system, and it was the same with that."[8] Doctors also feared they would develop cancer. Investigations then began, and these revealed that Allied too had dumped Kepone in the river. Upon learning of the poisoning and contamination, Virginia health authorities acted quickly to shut down Life Science and close the James River to fishing. Amid the "environmental decade" of the 1970s, Kepone pushed Virginia to approve a new Toxic Substances Information Act (TSIA) in 1976 and increase already existing momentum for greater state oversight of the environment. Court cases followed, including a landmark federal decision by U.S. district court judge Robert J. Merhige Jr. against Allied for its role in the disaster. The events associated with Kepone helped ensure passage of the federal Toxic Substances Control Act that same year. Two years later, the EPA banned Kepone for use in the United States.

The story continues beyond the 1970s and beyond Hopewell. Follow-up medical investigations on some workers in 1995 showed no signs of cancer, and the sterility and "Kepone shakes" dissipated. Around twenty thousand to forty thousand pounds of Kepone remain in the bottom of the James, buried under a few feet of sediment. The health of the river is better than it was in the 1970s, even as Kepone still shows up in fish samples, albeit at levels well below what is considered safe by federal regulations. While Virginia avoided a worse fate, the same can't be said elsewhere. There is a critical public health crisis on the islands of Guadeloupe and Martinique, where chlordecone's heavy use against the banana root borer poisoned the land, drinking water, and fishing stocks. The events in Virginia and in the Caribbean led the United Nations to ban the pesticide in 2009 under the provisions of the Stockholm Convention on POPs, persistent organic pollutants.

On the one hand, this story of an environmental disaster is a sadly predictable one of toxic poisoning in a southern community. Investigations revealed aspects of this environmental tragedy that are all too common: corporate avarice, ignorance, and regulatory failure. Calls for retribution invoked elements of environmental justice and led to vigorous debates over the science of toxicity and cancer. But for all of Hopewell's similarities to well-known stories from America's toxic twentieth century, this story offers some surprises as well as new insights and important connections among its many historical lessons. Fortunately, no workers died directly from Kepone exposure, though many workers at the Hopewell plant experienced unpredictable tremors and pain. With medical intervention, they were able to depurate all or virtually all of the Kepone.

No study has yet been conducted to measure the incidence of cancers among Kepone workers. Hopewell, though, ranks among the top polluters in Virginia and general incidences of cancer. Most workers were spared a worse fate because of another factor rarely present in southern environmental histories, active state intervention. While some officials dragged their feet, several government authorities as well as doctors and scientists acted boldly to stop further Kepone poisoning, discover a way to treat the workers and address environmental contamination. While Kepone still shows up in some fish species, levels have dropped significantly since the 1970s. The crisis led to new state and federal laws regarding toxic substances. Lawsuits brought some relief to workers and the seafood industry, and they also resulted in an endowment that continues to make a positive impact on Virginia's environment today.

By following Kepone's connections we can also extend the story beyond Hopewell. While the book maintains its focus on the city and its people, Kepone did leave Hopewell and found use on faraway fields where growers used it to combat insects. On the Caribbean islands of Guadeloupe and Martinique, workers spread the white powder on banana plantations, little aware of the crisis to emerge from the poison accumulating in their bodies and the soil and water around them. The islands shared some of the same experiences with Hopewell. There were negligent state officials and irresponsible corporate actions, as well as uncertainty and ignorance among regulators. Workers and residents both suffered from Kepone poisoning, and the compound damaged the environment. After the crisis, legal and regulatory measures ensued. Yet there are important differences as well. Hopewell's workers were white, working-class men who mostly got better, and for them some elements of the state mobilized to help. Hopewellians were spared widespread poisoning as well. Kepone remains in the James, but the river is healthier. In the Caribbean, chlordecone reflected a long history of racism and colonialism, and the French government in Paris favored growers and ignored the islanders' pleas for help for some two decades. Tragically, the poison contaminated large swaths of soil and water that affected the wider population. With the soil composition varying, the half-life of chlordecone on the islands ranges from several decades to several centuries.⁹ The Caribbean connection shows us that the Kepone story in Hopewell could have been worse.

There's much to be gained from this story—not just in what went wrong but also in what went right and the unexpected outcomes and silver linings. Understanding the Kepone story in Hopewell is critical to those who continue to struggle with chlordecone poisoning in the French Caribbean and for how we manage chemicals and their human and environmental impact now and in the future. To tell this complex tale of poisoning and recovery of both people and nature, I use the concept of "residues" as a connecting thread throughout the book. The idea comes from scholarship in science and technology studies to understand the temporal and spatial extent of chemical relations—the many connections that evolve from their production, use, disposal, monitoring, and regulation. Residues point to not only the molecular and biological remains of chemicals in human and natural environments but also their psychological, regulatory, and legal remains. We can move across time and space, from the molecular to the global, to track the past, present, and future of Kepone to better understand pesticides as part of the massive planetary terraforming that marks the Anthropocene, a term used to describe the most recent geological epoch in which human beings are the primary cause of planetary change.[10]

The Kepone story draws on other threads of scholarship as well. This untold history of Kepone both affirms and complicates the nexus between toxic disasters and the South's lax regulatory history. This study adds to those works that detail the struggles of everyday citizens against corporations and often irresponsive experts and government agencies as they face a health crisis.[11] Yet this narrative offers some unexpected twists and turns that set it apart from some of these works. While the response from authorities was uneven, portions of the state did act to end the crisis and address its effects. In part, this reflected the overlap of a broader shift in environmental awareness in the 1970s coupled with a change in Virginia politics as the state began to veer away from the days of the Byrd Machine, the Democratic organization run by the governor and then U.S. senator Harry F. Byrd from the 1920s to the 1960s. With political labels in flux, there was room for environmental action from so-called progressive southern governors like Virginia's Linwood Holton and Mills Godwin Jr. One factor may have been the reapportionment of voting after court decisions overturned district boundaries that for a long time gave rural areas more power than urban areas. These changes allowed growing numbers of urban and suburban voters to flex their environmental muscles in other southern states, like Maryland and Florida, and this may have been the case with Godwin as well as Holton. Virginians also approved a new state constitution in 1970 that mandated protection of the environment.[12]

Thinking spatially, the book links local environmental history to Virginia, the modern South, the nation, and the world. This account follows environmental historian Paul Sutter's appeal for scholars to "at once embody and transcend the region," situating it in a "larger world of connections," and to "reach out to the rest of the world."[13] While the book is not a global history of chlordecone, it does show Hopewell's connections to the wider world. Kepone allows us to look outward from the Life Science factory and the bodies of workers in Hopewell to the soils, waters, and bodies of residents in Guadeloupe and Martinique. This opens vistas into the global chemical relations of pesticides.[14] Kepone and other pesticides represented another aspect of the postwar American Century, symbols and tools of U.S. and corporate global power. With sales around the world, Kepone served agriculture and development projects, asserting both "technological modernity and ecological modernity."[15] This was tied to U.S. expansion of power in the Cold War, with the underlying belief that U.S. commercial expansion continued to be "a great work of peace, a noble cause."[16] It acted not only as an effective pesticide but also as a vector to reinforce a host of sociopolitical and cultural expectations surrounding insects and agriculture.[17] Of course, many of these chemical creations shifted to military applications, especially things like DDT and Agent Orange.[18] The domestic market reflected the growing reliance on chemicals as well, with pesticide use for homes and farms skyrocketing after World War II.[19]

In addition, so much environmental history of the New South examines agriculture, and this book certainly shows how the importance of farming led Hopewell into the chemical industry. But this book shifts the historical gaze toward understanding and documenting the growth of the modern South's environmental management state in the 1960s and 1970s, especially in areas of toxic substances and monitoring of the marine habitat outside the Chesapeake Bay and Atlantic Ocean.[20] "Environmental management state" is a term that refers to the line of scholarly inquiry devoted to studying the relationship between the state and the management of environmental issues. It is akin to concepts such as the welfare state or national security state. Deployed by environmental historian Adam Rome regarding the United States in the twentieth century, it is a term used often in the study of governments and societies throughout history and their relationship with nature.[21] Further, tracking the Kepone crisis entails coming to terms at some level with the notion of southern exceptionalism. Kepone showed that at least some local and state leaders could embrace aspects of environmentalism in the 1970s, bringing Virginia and Hopewell into the larger national trend. While the South has a strong share of toxic sacrifice zones and fenceline communities adjacent to polluters, cases of environmental contamination and poisoned workers span the nation and the globe.[22] Moreover, Hopewell's chemical and industrial connections, along with its ethnically and racially diverse population, reflected the broader trend of southern modernization and globalization in the twentieth century.[23]

Following Kepone's chemical connections requires examining the effects of poisoning on the bodies of workers and residents in both Hopewell and the Caribbean. These aspects of the story remind us of the costs for local people and communities in the global dependency on pesticides. Workers in Hopewell and residents of Guadeloupe and Martinique invoked biocitizenship, claiming rights based on harm to their health.[24] Workers in Hopewell found some restitution in the legal system, and their voices helped generate support for new laws. They and others made their case based on their masculinity, Kepone threatening both their working capacity and their biological capacity to reproduce. There was also a cruel irony in their fate; whiteness in Hopewell dovetailed with their masculinity to put them in harm's way, as the jobs they acquired came through formal and informal networks of labor that excluded women and African American men.[25] Further appeals to work and masculinity came from watermen, the men and women who make their living by fishing, crabbing, and oystering. Kepone contamination threatened their livelihoods. In the Caribbean, the people most affected by chlordecone are those descended from enslaved people. Men in the islands made similar arguments to those in Hopewell but were less successful in gaining legal restitution as chlordecone ravaged their way of life and their bodies. Moreover, the long-term toxicity in the islands affects women and children as well, through issues such as shorter gestation periods in pregnancy and premature births. In the islands, chlordecone evokes more directly the racism of the colonial past that still influences politics, labor systems, and the public health. The similarities and differences among each community's people reminds us of how uneven the costs of global terraforming are in the Anthropocene.

As much as the human element is central to the story, so too is the James River. It formed a dynamic interchange with human societies that developed alongside it, from Native Virginians through today. Ideas and uses of the river mixed with the river's own hydrology, ecology, and energy.[26] Fish and marine life like oysters played a key role in the Kepone story, not only because of their commercial value but also because their biology affected their uptake of Kepone and how officials and scientists responded to regulating the river. The knowledge of fish and oysters in the river became an area of conflict among state officials, scientists, political leaders, and watermen. Anglers and harvesters used their knowledge and influence in the political process to challenge state and federal officials over the fishing ban. Eventually Virginia lifted the ban, but warnings remain for consumers.[27] Concern over Kepone also reflected the postwar environmental shift toward valuing the ecology of the river and using a cleaner river as an amenity for attracting new types of development.

Inside these conflicts and those over how to assign accountability for the Kepone crisis swirled uncertainty, "a central and disconcerting feature of histories of toxicity."[28] Groups and individuals wielded it in ways depending on their interests. At times they did so intentionally, other times not. Scientific study of Kepone only began in earnest after the exposures in Hopewell brought the pesticide to a wider audience. In effect, the poisoning and contamination of the waterways brought Kepone exposure into existence, with scientists, regulators, workers, lawyers, watermen, and politicians all constructing knowledge on the fly in the face of a crisis.[29] Faced with the pressure and the uncertainty of Kepone's effects on marine life and humans, regulators from Virginia and the federal government closed the river to fishing and harvesting crabs and shellfish, and they enacted a set of action levels for Kepone in various marine species. Poisoned workers faced uncertainty in knowing their long-term health damage as they sought restitution in court. Armed with their own set of experts and data, watermen and their political allies used these uncertainties to challenge regulators for lower (or no) restrictions on which species they could catch and sell. Appeals to uncertainty and ignorance also came from Allied Chemical and Life Science as they fought off efforts to assign blame to them for poisoning workers and contaminating the river.[30]

Another key approach to the book is its personal focus. The stories of those who lived through it, experienced it, and made it happen in the way it did are the cornerstone of my retelling. I use oral history as a key source to ground the larger environmental story in the human one of actions, decisions, causes and effects, and lingering memory. Given the book's primary focus on Hopewell, I conducted more than twenty interviews with a variety of participants and observers of the Kepone crisis, including workers, scientists, and high-ranking public officials in the city and elsewhere to explore their own sense of the crisis, their actions, and their memories as to the importance of the events to them and to the larger world. In other parts of the book, where I could not do oral history, I looked for ways to bring the human element into the narrative, through archival records or newspaper stories. The personal focus connects to my own memory. As a child growing up along the James River, I remember hearing from adults about Kepone in the river and being careful about eating fish caught there. Although as a child I had little concept of the magnitude of the problem, I understood then it had something to do with poison in the water. This memory stayed with me and has been a major prompt for my interest in pursuing this topic.

The first two chapters tether the immediate contamination to the historical relationship between people and the James River, as well as the city of Hopewell, Allied, and its development of Kepone. The next three chapters address the government investigations that uncovered Kepone and examine the lawsuits related to the workers and to the water pollution in the James River that included debates over the toxicity of Kepone. Chapter six addresses how Kepone led to an expansion of the environmental management state, with new laws covering toxic substances. Chapter seven follows Kepone to the French Caribbean, where the pesticide looms even larger. It also returns to Hopewell to trace the legacy of Kepone in the city. Throughout, the book affirms how the story of Kepone opens new ways of understanding the development and impact of chemical production, use, disposal, monitoring, and regulation, with Hopewell at the center.

CHAPTER 1

The James River before Kepone

Rivers are not merely backdrops to history. The riverscape of the James reflects how river systems and human societies shape one another, forming a dynamic interchange between the human and nonhuman worlds. The Kepone contamination of the river can be understood as part of that interchange. Native people lived in a riverscape that reflected subsistence agriculture, hunting, gathering, and fishing, and use of the river for trade and transportation. Colonization by the English meant keeping the river for trade, transportation, and fishing but adding more intensive and extensive agriculture, including tobacco plantations, and a slave labor system. In the early nineteenth century, to assist with the advancement of the market economy and nation-building, Virginia and U.S. government officials, engineers, and business leaders changed the course of the river through canals and altered its flow with dams that harnessed the river's energy for industrial and residential use. The river had long been used for waste disposal, but post-Civil War industrialization and urbanization witnessed greater amounts and more toxic levels of waste entering the James, which began to damage the river's ecological health. Commercial overfishing and harvesting added to these pressures even as subsistence and recreational use of the river persisted. Local people fished, swam, and harvested oysters alongside these more commercialized uses. In the decades after World War II, individual and organized efforts to clean up the James emerged, becoming part of the postwar environmental movement, emphasizing ecological health and aesthetic appeal. Virginia's government responded to these developments, creating new laws and regulations along with federal ones as the environmental management state evolved. These laws and regulations would be important in dictating the state and federal response to Kepone.

These various human ideas and actions interacted with the dynamics of the river. Like all rivers, the James is itself an agent, a "provider of energy and resources, and a driving force" in history.[1] Hopewell's industrial development came in large part because of its direct access to the James River. Industries like Allied benefited from the river for shipping and the use of water for industrial supply and for dumping waste. The interactions went even deeper. In the case of Kepone, for example, the pesticide's movement through the river and impact on marine life depended on the river's ecology, water flow, and sediment patterns. Indeed, the biology and habits of various species in the river—each dependent on mixtures of salt water and fresh water, temperature, and water flow—shaped much of the human development of the river. These species are connected through the food web of the James, which meant that Kepone moved along the web in a process of biomagnification—greater concentrations at higher ends of the web.

There were multiple stakeholders involved in human society who were dependent on each species (and there still are). The watermen who caught the fish, crabs, and oysters were mostly independent small business owners, with a long tradition of working on the water. At the time of Kepone, there were some 8,000 commercial fishers across Virginia and around four hundred processing and wholesale operations that employed several thousand workers, most of whom worked seasonally. Of these, around 1,450 individuals made their living directly from the James River, with another 1,270 involved in processing. Many had long combined fishing or oystering with farming, yet by the 1970s this was a difficult combination to sustain, between the economic and ecological changes in fishing and the decline of economically viable family farms.[2] Beyond the watermen, there were the many marinas, boat builders, mechanics, and restaurants connected as well. There were also the thousands of individuals involved in recreational fishing and the businesses dependent on them. Given these connections, fluctuations in the harvest, up or down, could have a major economic impact throughout much of the state and beyond.[3]

These past uses and misuses of the river are the first residues of the Kepone story. They remind us that the material legacy of human interaction remains in and around the riverscape: dams, canal locks, slave cabins and plantation houses, palisades and gravesites, weirs, projectile points, bones of prehistoric megafauna, and discarded oyster shells. The stuff of more recent history accretes to the more distant residues: outflow pipes from city sewer systems and industries, railroad trestles, highway bridges, floodwalls, fishing trawlers, litter, and toxic waste like PCBs, mercury, and Kepone. The nonhuman world of the river offers its own set of residues, whether in the accretion of oyster shells into reefs or in the formation of sediments. These material items signify worldviews and social systems that left their own connections to the riverscape of Hopewell. Understanding the case of Kepone requires introducing some of the historical and ecological interactions and trends that shaped the riverscape, beginning with the river's formation.

Formation and Hydrology

Just south of the small town of Iron Gate, in the Appalachian Mountains of western Virginia, the Jackson and the Cowpasture Rivers come together to form the James River. From this point for 242 miles, the river winds its way to the Fall Line at Richmond. Here it becomes tidal and begins to widen as it flows another 106 miles until it empties into the Chesapeake Bay at Hampton Roads. The James is Virginia's largest river, with a watershed of some ten thousand square miles that touches around three million people, thirty-nine counties, and nineteen cities. It is the third-largest to empty into the Chesapeake Bay, the Susquehanna being first and the Potomac second. The river's watershed is in three sections. The Upper James begins in the Alleghany and Blue Ridge Mountains and flows to Lynchburg. The Middle James runs from Lynchburg to the Fall Line at Richmond. The Lower James runs from Richmond to the Chesapeake Bay and flows through Virginia's coastal plain.[4] In the tidal area of the Lower James, a wedge of fresh mountain water rushes downstream over a wedge of heavier salt water pushed upstream. Here, in an area from Claremont along the south bank down to Jamestown on the north bank, is the river's maximum turbidity zone, the cloudiest part of the river full of sediments. It's the place where most of the Kepone settled.[5]

Fish, Oysters, and Crabs

The food web in the James is key to understanding the impact of Kepone. The toxin settled into the sediment, persisting along the river bottom with various organisms that take in Kepone particles as they feed. In turn, larger organisms eat these smaller ones until eventually humans can consume seafood laced with toxic substances. Kepone bioaccumulates in each species, but consumers higher up the trophic level in the web face higher levels of concentration—biomagnification. Distinct marine species can also exhibit differing interactions with Kepone and accumulate or depurate the toxin depending on variables including individual biology and how long each species stays in the river. The great fear surrounding Kepone came from the potential economic and ecological harm the poison might unleash on highly valued species of fish, oysters, and crabs.

As a tidal river, the James possesses anadromous and catadromous fish. Anadromous fish are born in freshwater streams, mature in the ocean, and then return upriver to spawn. Main examples involved with Kepone include the American shad (*Alosa sapidissima*) (*sapidissima* being Latin for "most delicious") and striped bass (*Morone saxatilis*).[6] Catadromous fish are born in the ocean, mature in freshwater streams, and return downriver to spawn again in the sea. Of this type, the

American eel (*Anguilla rostrata*) is the main species involved with Kepone. For both types, evolutionary adaptations allowed these fishes to overcome various natural impediments to their spawning cycle. Issues such as water depth, velocity, temperature, and light intensity all influence the way each fish migrates and whether they reach their destination. Each fish also possesses unique characteristics in terms of biology and interactions with its marine ecology.

For example, all American eels are born in the Sargasso Sea south of Bermuda—although no one has witnessed the spawning secrets of the fish. The eels begin as larvae that drift with the currents for about a year toward the United States, the Caribbean, and Central and South America. They develop fins by the time they reach the coast and begin swimming upstream in rivers in late summer. In the James, like other estuaries, they live for many years (anywhere from two to eighteen), being active at night from spring to late summer, then burying themselves in the mud until spring returns.[7] In the James, eels seem to feed mainly on small crustaceans, insects, and crayfish. They also serve as prey for larger fish as well as fish-eating birds and humans. When mature, they return with millions of their counterparts to the Sargasso to spawn once, then die.[8] This isn't true for the anadromous striped bass (stripers or rockfish) and shad. Most stripers that come to the James migrate from New England to spawn. Striper larvae consume bottom-dwelling Mysid shrimp and plankton, moving on to small fish as they mature.[9] Shad and herring originally ascended the full length of the James River and entered the Jackson and Cowpasture Rivers near Covington. Shad were one of the most commercially valuable catches in the Chesapeake. They range from northern Florida to the Labrador Sea. Each spring, shad return to their home rivers after four or five years, swimming in schools in the river channel near the bottom to ascend the river to spawn.[10] Juvenile shad consume plankton and insect larvae while in the river. In late fall or early winter, they leave the river to return to the ocean to mature. Other predators in the food web eat the shad, including striped bass, bluefish, channel catfish, and humans.

Catfish became part of the Kepone story as well. By the time of the discovery of the Kepone contamination, there were four species in the James River. The white catfish (*Ameiurus catus*) is native to the river, and the channel catfish (*Ictalurus punctatus*) was introduced as a recreational fish in late nineteenth century. These were the two main types of catfish at the time revered for their recreational value, with some commercial use. Blue catfish (*Ictalurus furcatus*) were introduced by the State of Virginia into the James and Rappahannock Rivers in 1974 as part of the development of recreational fishing, and flathead catfish (*Pylodictus olivaris*) were introduced into the James accidentally in 1965 when a temporary holding pond at the Hog Island Game Refuge gave way during a storm and released about fifty fish. Although viewed as a source for greater and more lucrative recreational fishing tournaments, experts now see both the blue and the flathead as invasive species. Among other creatures, flathead catfish prey on shad and herring, threatening the latter's status. Blue catfish include menhaden and blue crabs in their diets, both important commercially as well as ecologically.[11]

Besides fish, oysters were a critical part of the Kepone story. The eastern oyster (*Crassostrea virginica*) is found along the estuaries of the eastern Americas, from the Gulf of St. Lawrence down to the Gulf of Mexico, and in the West Indies and coastal Brazil. It was introduced to Hawaii in 1866 and grows there too.[12] Oysters moved into the Chesapeake area around sixty-five hundred years ago as salt water, driven by rising oceans, pushed further into the bay and rivers. The eastern oyster is a bivalve with two rough shells that vary in color from white to gray or tan. It is a filter feeder, taking in phytoplankton and suspended detritus by opening its shells. Larger oysters can filter some fifty gallons of water per hour. They live in salty waters concentrated together on reefs, along rocks or oyster bars closer to shorelines, in water between eight and thirty-five feet deep. The reefs they build host several species of invertebrates and fish. Biologists describe them as "colonizers" and "ecosystem engineers." As colonizers, oysters are highly prolific, with a wide distribution of offspring. The eastern oyster is often the first species to occupy a suitable area for reproduction. As engineers, the oysters modify their physical environment for long-term survival by building their reefs. The concentration of oysters allows for genetic mixing between older and younger generations, conserving genetic diversity. Spawning occurs in the early summer as water temperatures rise. After two or three weeks, oyster larvae grow a "foot" that they use to crawl over the beds and find a location to settle. Once settled, the larvae secrete a glue-like substance to secure the left shell to the reef.[13]

Kepone affected another important species, blue crabs. They are usually associated with the Chesapeake Bay, but blue crabs are found in the Lower James. Crabs are bottom dwellers, omnivores that prefer mollusks but will feed on a variety of plants and marine life. They also serve as food for fish, eels, and other blue crabs as well as birds of prey and humans. They inhabit all areas of the water, with shallow grass and near-shore habitats serving mainly as nurseries and mating grounds, while mature males prefer deeper channels. Male crabs also venture farther up the river and generally prefer fresher water; females are the opposite and stay closer to the saltier portion nearer the bay, where they spawn, sending their young to the ocean to develop before they return to the bay and eventually upriver to grow into adults. The scientific name *Callinectes sapidus* means "delicious, beautiful swimmer," indicating perhaps why the creature is so popular, ending up as crab cakes or soup. Virginia's total in recent years was around $24 million in commercial value, and crabs are the most valuable fishery in the bay.[14]

Commercial Fishing since Industrialization

These species were critical to the seafood industry along the James River. Commercial interest in fish grew significantly after the Civil War, prompting federal and state authorities to intercede to study and try to regulate fishing and harvesting. The Kepone crisis built on these foundations. The United States Fish Commission formed in 1871, and Virginia followed in 1872 with the Virginia Commission of Fisheries, with the responsibility to stock and replenish food fishes like shad and stripers.[15] Another area of concern was oysters, both in the river and more so in Virginia's part of the Chesapeake Bay. Oyster populations in the Chesapeake remained strong until the early 1800s, when the collapse of the New England oyster fisheries pushed those industrial fishery vessels to the Chesapeake. This came with advances in industrial technology "including steam locomotion, canning, and food preservation," along with dredging for oysters (replacing the manual use of tongs).[16] The expansion led to the first observations of public oyster ground depletion in the late 1850s. Demand led the James River to become a leading location for planting "seed" oysters in the lower salinity reaches that could be transplanted elsewhere around the bay and even in the Northeast.[17] After a lull during the Civil War, harvesting boomed again, prompting the "oyster wars" of the 1870s and 1880s that witnessed conflicts among watermen and between them and the governments of Maryland and Virginia. Both states created oyster navies to enforce the laws, and each often engaged in running gun battles on the water with various watermen. The war came as harvests remained high through the 1880s. But sometime after 1890 they declined from the public grounds as reefs were depleted. The war subsided, but the relationships between and among watermen and the government continued, relationships that were sometimes cooperative, sometimes not, a key aspect to the Kepone story.[18]

Like oystering, commercial crabbing did not begin in earnest until after the Civil War, once railroads developed along the southern coast to ship crabs on ice to more distant markets. Canning of hard crab meat began in Hampton, Virginia, in 1878. Methods of crabbing also changed as supply and demand grew. Crabbers used the trotline to harvest crabs—"a long line with baits tied to the line" anchored at both ends. In trotline use, the crabber brings the line to the surface over a roller and then catches the crabs in a dip net. The catching of soft-shell crabs grew after 1870 when L. Cooper Dize of Crisfield, Maryland, invented the scrape, a "small toothless dredge" for harvesting them. The next major development came with the widespread development of the crab pot in the 1930s, today still the most widely used method, albeit with some modifications.[19]

Pollution before Kepone

Commercial fishing grew alongside industrialization and urbanization, increasing the use of the river as a waste destination. The Kepone that came from Hopewell was one part of a longer trend in the James and its tributaries becoming depositories for waste, residues of modernization along America's founding river. Beginning in central Virginia, Richmond's footprint expanded as it rebuilt after the Civil War. The growing use of electricity required larger amounts of water for cooling in power stations, water that, when discharged, created chemical and thermal pollution. With the expansion of flush toilets, the city constructed more and more pipes, which discharged untreated waste into the James.[20] Later Richmond and other cities along the river constructed combined sewer overflows. During heavy rainfall or snowmelt, these systems collect both city waste and storm runoff that bypasses treatment plants and dumps the water directly into the river. The city dumped a combined 44.7 million gallons of raw sewage and industrial waste into the river before adding a treatment plant in 1958. This improvement came only with federal help, as the federal Water Pollution Control Act amendment in 1956 provided grants to states and municipalities for construction of wastewater treatment plants.[21]

The urban and industrial expansion of areas south of Richmond, especially Hopewell and the Hampton Roads areas, added pressure and pollution into the James. Point discharges in Hopewell included the city's sewage system, as well as Fort Lee (an adjacent U.S. Army installation) and chemical industries like Dupont and then Allied. At the mouth of the river, the rapidly developing Hampton Roads area contributed extensively to water pollution. The combination of industrial, urban, and military installations such as the Norfolk Naval Base, Hampton's Langley Air Force Base, and the Newport News Shipyard placed new burdens on the river and its marine life, especially the vulnerable oyster beds nearby.

The environmental management state in place at the time of Kepone had roots in earlier developments over pollution. Virginia created several conservation-based government agencies between 1900 and 1930, including the Department of Forestry in 1914, the Department of Game and Inland Fisheries in 1916, and the State Commission on Conservation and Development in 1926. State leaders, especially Harry Byrd Sr., played a role in establishing the Shenandoah National Park in 1926.[22] Several state government reports documented studies of the growing water pollution as the twentieth century progressed and led the state to craft the 1946 Water Control Law.[23] There was, however, no central governmental agency responsible for water pollution or management of water resources generally, and the state's spending maintained a pay-as-you-go system. Still, these developments in environmental management were part of the developing mindset of

Virginia political leaders in the Byrd organization: "a conservative preference for traditional practices, especially when it came to spending money, versus a zest for progress, growth, and the utilization of resources in the most efficient and productive manner possible."[24] Byrd consolidated control over Virginia politics, supporting limited government and racial segregation, with a grip that would last until the 1960s. Meanwhile, the desire for "industrial development, highways, and tourism" drove the concerns with pollution as part of the general trend of 1920s "business progressivism" to make the state appealing as a destination for investment and tourism.[25]

As across the United States, the post-World War II era witnessed increased environmental concern, with the James a subject of study. The Virginia Academy of Science created a James River Project Committee that published *The James River Basin: Past, Present, and Future* in 1950. Highlighting the various activities in the basin, the book signaled growing public concern that the resources of the area would not be sufficient to support population growth. The committee wrote, "[Virginians] have used the resources of the Basin for great historic achievements and for the attainment of a high place among civilization's builders," but it added that the "practices of... past... management of our basic resources" would not suffice in the presence of growing population.[26] Concerns over the condition of waterways escalated in the 1960s. The Virginia Institute for Marine Science saw its responsibilities for research and monitoring increase, and the state legislature increased its capital outlay to the organization from around $500,000 in 1958 to over $3 million by 1972. Grants and contracts increased as well.[27]

A 1966 report on the tidal James by the Virginia Institute of Marine Science noted the river's many uses and its compromised health. The report noted that the tidal James was a multipurpose natural resource: "Its aesthetic value is of immeasurable importance to the millions of tourists who stand on the shores of Jamestown Island." It was also, the report noted, "a major oyster producing area," with over "500,000 bushels of shucking oysters" harvested and over "600,000 bushels of seed oysters" caught and replanted to other rivers. At the time of this report, the breeding populations of oysters were being ravaged by MSX, a disease caused by *Minchinia* (now *Haplosporidium*) *nelsoni*, a protozoan parasite that had entered the Delaware Bay in the 1950s and spread to Virginia by the 1960s.[28] The tidal James also supported "significant sport and commercial fin-fisheries," the report noted, including "striped bass, spot, flounder, croaker, trout, shad, and herring." Heavy industrial use also characterized the tidal area, especially from Richmond south to Hopewell, and Hopewell drew its water from the Appomattox River (which, as the report stated, "exchanges with James River water during each tidal cycle"). There was limited use of river water for agricultural irrigation, and the James was also "an important waterway for commerce and military activities" and the Hampton Roads area was (and still is) "a major port with shipping and ship building facilities." The river was also used for "domestic and industrial waste disposal," which the report said had led to "hundreds of acres of oyster beds" being condemned for direct harvesting. Algae blooms, fueled by excessive nutrients entering the river through these waste discharges, were becoming common in the summer. Low dissolved oxygen levels created dead zones in the river, one large notable area being below Hopewell, and fish kills were being recorded. In some areas of the tidal James "the bottom was completely devoid of animal life" because of toxic materials.[29]

The same year, the James River Basin Interleague Committee of the Virginia League of Women Voters published *James River Basin: A Progress Report*. The committee was rooted in a national League of Women Voters initiative to study and support conservation of water resources. The report noted serious pollution from industries in the upper reaches of James, particularly from paper mills and acid wastes from American Cyanamid Company, a chemical manufacturer just below Lynchburg. The report also noted heavy sewage and industrial pollution in the river around Hopewell. Despite a pollution control program coming into existence in Virginia in 1946 through the State Water Control Board, industries continued to pollute above acceptable levels, and convictions under the law for violations had been minimal.[30]

Richmonder Fitzgerald (Gerry) Bemiss led efforts in Virginia's General Assembly for increased state government attention to environmental issues and for greater attention to the James in particular. He chaired Virginia's Outdoor Recreation Study Commission (1964–66) that led to the establishment of the Commission of Outdoor Recreation and the Virginia Historic Landmarks Commission. He chaired the Special Committee on Water Resources, charged with determining the effects of growth on water resources. He was involved in Richmond-area conservation and planning issues as well. In 1976 he helped form and then served as president of the James River Association, a group of concerned citizens who lobbied for improved health of the river. The organization remains active today.[31]

These concerns and actions at the state level came along with the growing presence of the federal government in addressing water pollution and regulation. These laws would come to bear on the immediate response to Kepone. Until the post-World War II years, the only major federal law related to water pollution was the Rivers and Harbors Act of 1899, which made it a misdemeanor to discharge refuse matter into navigable waters, or tributaries of them, of the United States. The federal Water Pollution Control Act of 1948 created the first major presence of the federal government in clean water issues. It still left to the states the primary responsibility for maintaining clean water, but it encouraged water pollution control and provided technical assistance and funds to address problems. Between the

1950s and 1965, federal oversight increased, including 1956 amendments to provide money for wastewater treatment plants. Further amendments to the original law in 1965 required states to set standards for interstate waters that would then be used to determine pollution levels and control requirements. The 1965 changes reflected the shift from protecting not only human health through drinking water but also ecological health and uses including swimming and fishing.[32]

Political Changes

As new political constituencies were in play, the environment became a key issue and part of Virginia's changing political dynamic in the late 1960s and early 1970s. One of the leading political voices in the U.S. Senate on environmental issues was Virginia senator William B. Spong Jr. The moderate Democrat managed to defeat the conservative A. Willis Robertson in the 1966 Democratic primary election and go on to win the general election that year. Spong's victory was a major blow toward ending the political machine founded by Harry F. Byrd that had run Virginia's government since the 1920s. Spong would support several environmental causes, especially pollution control, helping to author the Clean Air Act of 1970.[33]

Adaptation to a new political world defined two governors central to the era's environmental concerns, Mills Godwin Jr. and Linwood Holton. Both came from an era when postwar environmentalism was "broad enough in its ideas and values" that it could "accommodate a wide variety of political philosophies." It could cut across ideological boundaries so that Republicans and Democrats could embrace it.[34] This came through in Virginia. With party labels in flux, the policies associated with the environmental decade of the 1970s could find fertile ground under these two governors. A new political context emerged in the form of reapportionment after 1965 that opened the way to urban and suburban voters to influence politics with greater concern for environmental issues. In 1970 Virginia voters approved a new constitution (effective 1971), and among the changes was a new article that proclaimed the commonwealth's duty to "to conserve, develop, and utilize its natural resources" and to "protect its atmosphere, lands, and waters from pollution, impairment, or destruction."[35]

Elected as a Democrat in 1965, Godwin, like many southern politicians in the era, "supported federal intervention in his state when it created economic growth" yet tended to resist further encroachment. Earlier in his career he supported the Massive Resistance campaign against desegregation but toned down his rhetoric by the time he became governor.[36] He acceded to federal law and Supreme Court decisions that forced Virginia to revise its state constitution to remove barriers like the poll tax and to alter reapportionment in districts to remove the bias in favor of rural districts. At the same time, he sought to create a pro-business image for Virginia and opposed much of the liberal agenda associated with Lyndon Johnson's Great Society programs. He did orchestrate new investments in education, highway construction, and industrial development as well as increases in taxes to pay for expanding some state government services, including environmental protections.[37] After a term out of office, Godwin was elected for another term in 1973, this time as a Republican. It was during this second term that he faced the Kepone crisis.[38]

In between Godwin's terms, Linwood Holton made history in 1970 by becoming Virginia's first Republican governor since Reconstruction. He was more progressive than Godwin on racial matters, having opposed Democratic resistance to school desegregation. He also spoke in favor of greater environmental protections in Virginia. As he noted in his memoir: "There was no instrument of the state government that comprehensively dealt with the environment when I took office. I created one."[39] This was the Council on the Environment that included Gerry Bemiss as well as Jerry McCarthy, who would go on to lead the Virginia Environmental Endowment upon its creation following the Kepone disaster. One of the council's initial actions was a public listening tour in early 1971, visiting Richmond, Norfolk, Roanoke, Charlottesville, and northern Virginia. The council published a summary of these findings in a report, "Our Common Wealth: Virginians View Their Environment." Speakers raised several prominent issues and submitted statements on topics that included damage from highway development, increasing environmental education in the state, preservation of open spaces, pollution, overpopulation, and pesticides.[40] McCarthy recalled that the list of issues formed "an agenda for the next 45 years and counting."[41]

In 1970 came Earth Day (which Holton publicly endorsed) and the creation of the Environmental Protection Agency. Among other things, the EPA assumed responsibility for water pollution. In 1972, Congress approved significant amendments to the 1948 Water Pollution Control Act; as amended, the law became commonly known as the Clean Water Act.

These changes called for the restoration and maintenance of the chemical, physical, and biological integrity of the nation's waters. The amendments focused on identifying point sources of pollution, requiring all municipal and industrial wastewater to be treated before being discharged into waterways, and provided additional federal monetary assistance for municipal wastewater treatment. Virginia moved to build or upgrade new facilities under this program, and Hopewell was busy building its new treatment plant as the Kepone contamination shut down the digesters on the old one. The digesters used bacteria to break down organic matter in wastewater solids. According to Holton, he had planned to use an excise tax on tobacco to raise the needed state matching funds. "I couldn't get tobacco tax, but I scared Philip Morris so bad, they said, 'If you lay off our product, and not try to put an excise tax on us next time, we'll support a general tax increase.' They got the Chamber of Commerce to endorse an increase in the income tax and that's where we got the money to clean up the rivers."[42] The 1972 amendments also created the National Pollutant Discharge Elimination System to establish a procedure for approving permits to discharge pollutants into waterways. It would be these regulations that would come into play during the legal proceedings against Allied Chemical, Life Science Products, and Hopewell stemming from the Kepone contamination.

The Virginia General Assembly approved the state's Environmental Quality Act in 1972, which declared protection of the state's environment a priority and required consideration of environmental impacts in the decisions of state agencies. This followed the changes to the Virginia constitution. Later legislation in 1973 added a requirement of an environmental impact report by state agencies on projects over $100,000. The assembly also strengthened other aspects of environmental law in areas such as air and water pollution and protection of wetlands. Despite the wishes of Jerry McCarthy and other officials and environmental groups, the assembly did not create a state environmental protection agency—that did not come until the formation of the Department of Environmental Quality in 1993.[43]

These developments of the environmental management state reflected the growing concerns of the 1970s related to pollution. In particular, the state's rivers garnered a significant amount of attention, as these polluted streams were important as sources of commercial and recreational activities, amenities for new development, and water for growing cities and new industries. They often became symbols of the environmental crisis. The Virginia State Water Control Board assisted with funding for research, along with the Water Resources Research Center at Virginia Tech, funded through the federal Water Resources Research Act of 1964.[44] In addressing the polluted James, attention focused on key industries, military installations, and cities sending waste into the water. From the river, we now turn to focus more clearly on the immediate setting for the Kepone story, Hopewell—once the self-proclaimed "Chemical Capital of the South."

CHAPTER 2

Hopewell, Allied, and the Beetle Battles

At some point in the 1950s, Hopewell's city leaders dubbed their home the "Chemical Capital of the South." Marketing brochures and signs posted along main roads leading into the city advertised this claim, and chemicals became part of the city's identity. Former city manager Clint Strong understood this. "There was a lot of odor in the air, and people were saying 'Oh, that odor, it stinks.' Hopewell would say, 'That's money. What you're smelling is money.'" He went on to recall: "This was the attitude. Good pay. That attitude was a big thing, because plants were kind of great jobs; don't mess with the plants."[1] Linda Crowe grew up Hopewell. "We were a proud Allied Chemical family. My father was an engineer for the company. My mother occasionally took part time work there as a secretary. I worked in the purchasing department during breaks from college. My brother and many of his friends hired on as shift workers at Allied to make money when high school was out for the summer. My brothers and sister and I beamed with pride when we drove over the bridge, past the sign that bragged 'Welcome to Hopewell: Chemical Capital of the South.'"[2] Jeanie LeNoir Langford, Hopewell native, local librarian, and historian concurs. "It did have its faults and its flaws, and it occasionally smells a little rough, but that's just part of us being us. That's what the people here know. So to us, it's like, well, the wind needs to shift, but it's all good, and it's like I tell people, don't mind, I only glow in the dark after midnight, and it's just something you get used to, and that's all, for the most part, generation after generation that we have now."[3]

These recollections showcase a central element in how some in Hopewell saw the city. They recognized the environmental and health concerns associated with chemical production but at the same time suggest acceptance of the city's identity. Some bristle at the continued identification of Hopewell with Kepone, especially now that the James River and the creeks in Hopewell are much cleaner than they were. Wayne Walton served on Hopewell city council from 2008 to 2016, was vice mayor from 2010 to 2012, and is the former chair of the Friends of the Lower Appomattox River (FOLAR), a nonprofit formed in 2000 to conserve and protect the river's lower reaches as it joins the James. He worked for Hercules for forty--

Hopewell, Chemical Capital of the South. Visitors arriving to Hopewell along Route 10 in 1975 would have seen this sign boasting of Hopewell's chemical industry. The city removed the sign after the Kepone disaster. Virginian-Pilot Archives/TCA (Tribune Content Agency)

Some of Hopewell's factories, looking east. This photo, taken in January 1976, shows a portion of the sprawling Allied Chemical plant in the foreground. AP photo, by Mary Ann Carter

one years before retiring. In 2016 he said, "We get a bad rap, maybe rightly so, but it's forty years ago. Look at what's happened since then."⁴ Mark Haley, former city manager, expressed the same frustration. "[For many years,] no matter what national show I'd go to with the Water Environment Federation, everybody'd see my name badge. As soon as they saw Hopewell, yeah, Kepone. You'll still hear people. Not too long ago somebody threw that pie right in my face again. I said, 'Damn, it's been forty years. When are you going to give us a break? It's long over.'"⁵

As I interviewed residents in Hopewell, they often asked with a sigh why I was doing this book. Yet the residues of Kepone remain strong in the public memory, and they are woven into Hopewell's identity despite what some might prefer. Part of being a chemical capital means accepting the history of Kepone. It also means regularly showing up in lists of places and companies tagged for toxic releases and violations of environmental regulations including the Clean Air and Clean Water Acts. This dual sense of identity speaks to what historians Gregg Mitman, Michelle Murphy, and Christopher Sellers have said about historical approaches to the intersection between environment and health. "Economic regimes of accumulation and production," they write, "coalesce in particular locations and seep into the character, sensibility, and experience of everyday life."[6] This fits Hopewell. Rather than simply an indicator of an isolated event, Kepone is central to the history of Hopewell. Hopewell is a prime example of what Michelle Mart writes, that the American commitment to pesticides was more than material or health-related, it was cultural too. She calls this commitment to pesticides and other chemicals a love story, a relationship able to strengthen over time despite "accidents, controversies, rising expense, and declining effectiveness."[7]

Understanding how and why chemists with Allied developed Kepone and then manufactured it in Hopewell also demands some understanding of the post–World War II surge in pesticide use and the history of the two insects against which Kepone was deployed, the Colorado potato beetle and the banana weevil. In a postwar American Century, amid U.S. global dominance and the Cold War, the chemical industry surged forward to produce new formulations to fight pests at home and abroad but also defeat the evils of communism. Within this larger geopolitical context, both insects illustrate the ways in which culture and nature intertwine. Each adapted and evolved with the plants and the insecticides—including Kepone—designed to kill them. Each was, and remains, part of the human-natural system associated with potatoes and bananas. Over time, these pesticides can lose effectiveness, and even when they work the success is impermanent: there is no complete eradication. Kepone was but another contributor to the global pesticide treadmill that continues to characterize agriculture in the United States and around the world. Allied's global presence meant the company could sell its products like Kepone most anywhere. Kepone's use against the two insects—especially in Europe and the Caribbean—linked people and environments in those places with Hopewell, workers at Allied and Life Science, and the polluted riverscape along the James River. The residues of Kepone not only remain in the soils and sediments of Hopewell's environment, and the bodies and minds of its residents and workers, but also in the landscapes, bodies, and minds of workers and residents elsewhere, with devastating consequences in Guadeloupe and Martinique.

Hopewell's Early History

As they were doing at other points along the river, prehistoric people lived along the south bank of the James River in what is today Hopewell. Weyanoke Indians, part of the Powhatan chiefdom, occupied the land at the time of English settlement. Across the Appomattox River were the Appomattox Indians. Apparently the first English to see the land in today's Hopewell arrived in May 1607, part of Christopher Newport's group exploring the area for a settlement. They chose Jamestown instead and so left the area; had they selected Hopewell, who knows how things may have turned out. English settlers returned in 1613 when Sir Thomas Dale, then deputy governor of Virginia, laid out a plantation around the town of "Bermuda Cittie" at the confluence of the James and Appomattox Rivers. Later changed to Charles City, the name became Charles City Point and then simply City Point. Dale populated the settlement with indentured servants who completed their terms of service in 1617.

Adjoining the city were the Appomattox and Hopewell Farms, plantations created by Captain Francis Eppes in 1635.[8] Both City Point and the plantations became part of the newly created Prince George County in 1703. Although Edmund Archer's mural in the Hopewell Post Office shows a friendly meeting between Eppes and Appomattox Indians, Eppes had been in the area since 1620 and led assaults against the Appomattox and Weyanoke peoples during the Indian wars in the decade following. He returned to England to collect his inheritance and came back to Virginia in 1631 with his family and thirty indentured servants. It was on this return that he established the plantations from land granted to him by Virginia in agreement with England. These included those near City Point as well Bermuda Hundred and Eppes Island just across the James. By 1700 City Point was a central location for shipping and warehouses, particularly for tobacco and enslaved people. It was an important node in the transatlantic system of trade, building the global connections to the area that would become Hopewell.[9]

The Eppes estate eventually came under the control of Richard Eppes (b. 1824) from 1851 until his death in 1896. Like other planters, Eppes constructed a system that reflected his vision of order and control of nature, including the land, the river, and other people. During the height of slavery before the Civil War, he owned some two thousand acres and as many as 130 enslaved people.[10] Eppes was part of a new wave of plantation owners and agricultural enthusiasts who began to make land "improvement" synonymous with scientific knowledge, applications of machinery and other technical developments to the land, and use of fertilizers, the foundation for later chemical applications to crops and soil. Unknowingly prescient, Eppes wrote that "a farm is another name for a chemical laboratory."[11] Eppes serves as an important link in many of the interconnections in Hopewell's history relevant to Kepone. Through him came the deepening dependence on slave agriculture as well as scientific management that helped usher in the world of chemical applications such as fertilizer and pesticides.

Railroads came to City Point in 1838, connecting the town and the James River to Petersburg. These rail connections made City Point a valuable location during the Civil War; it served as General Ulysses S. Grant's headquarters during the siege of Petersburg. Following the Civil War, agriculture resumed its dominant place in the region around City Point, but planters—now using free labor as opposed to slave—added peanuts as a cash crop. Yet the soils did not hold up well, and by the time of World War I production had dropped, and many lands had been abandoned.

The Toughest Town North of Hell

Into this riverscape the city of Hopewell came into being with a violent rush. It began when DuPont purchased land along the James River from the Eppes family and constructed a company town and dynamite plant there in 1912. The Eppes family asked that DuPont name the town Hopewell after the former plantation.[12] The site and the plant together reveal the connections between chemistry, war, and agriculture so prevalent in Hopewell's history and the Kepone story. DuPont's interest in building the Hopewell plant came from dynamite's use in clearing land for agriculture in the South. The number of farms grew in the early twentieth century as more land came under cultivation, perhaps prompting DuPont to produce dynamite in Hopewell.[13] DuPont sold its 1910 *Hand Book of Explosives for Farmers, Planters, Ranchers* to instruct readers in the best use of dynamite for clearing land of stumps, boulders, trees, and other objects. They even held stump blasting demonstrations, including one in Virginia.[14] A DuPont advertisement in *Hardware Dealers' Magazine* in June 1919, directed to agricultural equipment manufacturers and dealers, noted that the South offered "a practically unlimited field for sale of tractors, riding plows, cultivators, reapers, mowers, [and] binders" but said that sales were limited because less than 25 percent of the South's cultivated land was free of stumps. Using dynamite, the ad said, could open more land for agriculture.[15]

With the start of World War I, what was a modest plant rapidly became the world's largest guncotton (nitrocellulose) facility. While the United States remained neutral when war began in August of 1914, DuPont signed contracts with the Allies of France, Britain, and Russia to supply them with guncotton, a component in smokeless powder and artillery ammunition. The contracts meant the rapid expansion of the factory and the creation of Hopewell, seemingly overnight.[16] The scale of new plant construction in 1914–16 and growth of the city is remarkable. The plant was to produce one million pounds of guncotton per day, and at its peak it employed some twenty-five thousand people working with 425 rail cars of materiel delivered every day. The plant and city used seventy million gallons of filtered water per day (enough to supply Boston) and burned ninety railroad cars of coal per day.[17] With plant construction underway, DuPont purchased an additional seven hundred acres of land to create a small city for the onslaught of people coming to Hopewell, a polyglot mix of local and regional whites and blacks, as well as first- and second-generation immigrants from Europe, including many from Russia, Greece, France, and Romania. In December 1915 a fire swept through much of the business district, and many areas of homes were also destroyed. Miraculously, no one died. The area was rebuilt in 1916, and the city charter adopted in July that year officially created Hopewell.[18] Hopewell resident A. V. Carey published *Pictorial History of Hopewell, Virginia* in 1916, which would be added to and republished in 1962.[19] In it he included the lyrics to a song, "Powder Town Rag," that included pejoratives common in the early twentieth century:

> Money flows quite freely here, 'tis no stingy town,
> The wages paid by Du Pont have won for them renown;
> The North and South, the East and West, all are here to seek
> A fortune from the Du Pont works—White, Dago, Wapp, and Greek.[20]

DuPont constructed 3,109 housing units in three housing areas. A-Village in the still separate City Point consisted of permanent housing for male company officials. For its male workers, DuPont built B-Village for whites and, in keeping with segregation, South B-Village for blacks. Initially, the apartments and bungalows in all of B-Village had tarpaper walls. Streets were dirt, and sidewalks consisted of wooden planks. Meanwhile A-Village had concrete roads and sidewalks at the start. The company also constructed several other facilities, including a railroad station, a clubhouse for supervisors, two hotels (for single men), two schools (one for whites, one for blacks), two YMCAs, and two company stores.[21] To combat malaria, the company drained streams and ponds and spread oil on standing pools of water.[22] Those who could not find housing in Hopewell commuted by trains from Petersburg and Richmond, and these commuters numbered perhaps as many as four thousand per day.

In addition, the War Department authorized the construction of a nearby military facility, Camp Lee, which opened in 1917. More than sixty thousand doughboys trained there prior to their arrival on the Western Front. The facility essentially became an unofficial city that added to the dynamism and chaos of Hopewell. In addition, during the influenza epidemic of 1918, Camp Lee's hospital treated some ten thousand infected soldiers there; some seven hundred died.

Hopewell's rowdy reputation developed early. Reports stated that Hopewell was akin to a "breech born, squalling infant, stump muddy and soot-smeared," that "sold beer, whiskey and feminine favors." A letter to then–governor Henry Carter Stuart in July 1915 reported lawlessness and vice running rampant in "our New City." "Women of questionable character from every city of Va. And other states," the writer said, "lure these young men away from their homes to destruction and death and Cabaret Shows." L. R. Driver, a newspaper reporter sent by Stuart to investigate the reports, called Hopewell the "toughest town north of Hell."[23]

Carey's Hopewell of early 1916 differed remarkably from the rowdy one reported elsewhere. With Governor Stuart intervening, Carey wrote that "law and order prevail[ed] everywhere" and that it was a "clean city, as far as morals go," with the "class of men employed" being "far above the average."[24] Perhaps things had become quieter for a period after the fire and as the city was being rebuilt. Yet reports from later in the year indicate Hopewell's rough reputation continued. In November 1916 Virginia enacted prohibition, but that didn't seem to alter things. Attorney General John Garland Pollard warned Governor Stuart about crime, vice, and illegal beer selling; there were "200 saloons in operation," according to Pollard.[25] Prostitution, drug traffic, gambling, bribery, and corruption were also commonplace.

Allied Chemical Arrives

With the armistice of 1918, Hopewell collapsed even faster than it had grown. Camp Lee closed, DuPont shuttered its plant, and the population dissolved to less than one thousand. Workers "abandoned the city as though it were besieged." "Whole sections of the sprawling city were left desolate, and silence reigned where life had teemed only a few days before."[26] Local boosters and DuPont then worked to revive Hopewell as an industrial city. Ignoring the recent past, or perhaps challenging it, DuPont printed advertising material to persuade businesses to locate on the former site of its facility. One piece was titled "An Answer to the Question of Labor Troubles and Production" and advertised Hopewell's favorable climate, access to water, passive laborers, and well-maintained housing, including "hundreds of bungalows, cottages and houses, with lawns and gardens bordering on well-paved streets."[27]

The efforts proved successful. First, Stamsocott Company maintained the connection to explosives when it opened a cellulose manufacturing facility in 1920 (later purchased by Hercules Powder Company) that expanded along with the uses of cellulose. The next major firm to come was the Belgian corporation Fabrique de Soie Artificielle de Tubize (incorporated in Delaware in 1920 as the Tubize—pronounced TOO-bees—Artificial Silk Company of America), which purchased parts of the former DuPont facility and began making rayon. Jeanie Langford, whose father would work as a chemical engineer at Allied, noted that the processes of making guncotton and rayon are strikingly similar. Tubize, she said, "made rayon from almost the identical process that the DuPont plant that made guncotton, was almost the identical nitrocellulose process, except a little bit longer in the acid, you get guncotton; a little bit less, you get rayon."[28] By 1928 it employed about thirty-five hundred people to produce ten million pounds of rayon annually. It was common in the 1920s for companies like Tubize to practice welfare capitalism, sponsoring a range of worker activities such as clubs and sports teams. Tubize became famous for one of these, its Hawaiian Orchestra.[29] Janice Denton, who was the office manager for Life Science Products, remembers that her father was one of those attracted to Hopewell for the jobs. "I think when my dad came, it was during the Depression, and word got around fast where they were hiring. He paid $700 for the village house, and today it's still there, right downtown."[30]

It was during the 1920s that Allied Chemical arrived in Hopewell. It began as Allied Chemical and Dye Corporation in 1920, the result of a merger of several companies: the Barrett Company, which supplied coal tar chemicals and roofing; General Chemical, maker of industrial acids; National Aniline and Chemical Company, makers of dyes; Semet-Solvay Company, which manufactured coke and by-products; and Solvay Process Company, producer of alkalis and nitrogen materials. They merged in the wake of World War I, during which supplies of dyes, pharmaceuticals, and other products had been cut off by Germany, then the world's leader in chemical production. The founders were determined to end foreign domination and strengthen domestic chemical production.[31]

In 1921 Allied created the first commercial nitrogen fixation plant in the United States, in Syracuse, New York. This provided the experience to then launch Allied's facility in Hopewell, which began production in 1928.[32] The plant made synthetic ammonia and other nitrogenous fertilizer solutions along with chlorine.[33] It became the largest producer of ammonia in the world and led to Allied creating a nitrogen division and additional plants in Ohio and Nebraska. From its beginning the company possessed prominent political connections. For example, Eugene Meyer, key architect of the merger that created Allied, served in President Woodrow Wilson's cabinet as head of the War Finance Corporation. He went on to lead the Federal Reserve Board from 1930 to 1933, became publisher of the *Washington Post*, and then became the first president of the World Bank in 1946.

The Great Depression brought a turning point in the history of Hopewell, with direct bearing on Kepone. Unemployment rates went up, of course, but those still employed in larger industries received support from the federal government under the New Deal of President Franklin D. Roosevelt to join unions, and they did so in large numbers, including in Hopewell. Tubize workers organized under the United Textile Workers. The company then fired five hundred workers for suspected union activity, and in June 1934 the eighteen hundred employees remaining joined other textile workers in a massive strike of some four hundred thousand workers across the nation. In Hopewell the strike failed. Tubize closed the plant in July. It reopened a much smaller section for knitting and dyeing that employed a few hundred workers, mostly women, who were not unionized. Some in Hopewell believe that Tubize may have been losing money at its Hopewell operation and used the strike as an excuse to close the plant and shift operations elsewhere where labor was cheaper and less prone to organizing. The Tubize strike holds a place in the collective memory of Hopewell for other reasons. It was a turning point that remains in the cultural residue of the city. For Kit Weigel of the *Hopewell News*, the strike had direct bearing on the response to Kepone. She wrote in the newspaper's 1976 bicentennial issue: "When Tubize left, she was not Hopewell's only industry. Her leaving, however, seems to have scarred the city's psyche and instilled a fear of strikes and of badgering an industry with environmental concerns until it pulls up stakes and settles where folks don't complain so much."[34]

World War II did not bring the boom town experience of the previous conflict, but it certainly allowed Hopewell's industries to prosper and build further connections among chemicals, agriculture, textiles, and war. Camp Lee reopened in 1941 and peaked at thirty-five thousand people in 1944. The army renamed it Fort Lee in 1950, and it has been a permanent fixture in the area since. Allied made sulfuric acid and synthetic nitrates used for explosives in World War II.[35] Hercules Powder made explosives as well. According to former employee Reed Belden, after the war Allied was part of a team that studied German nylon manufacturing and brought back information on processing that went into Allied's decision to manufacture caprolactam, a crystalline compound derived from ammonia and used in making synthetic fibers like nylon.[36] The caprolactam plant went to Hopewell to take advantage of the ammonia already being produced there and the deepwater access that the James River offered for supply of materials used in manufacturing.

In having a vibrant chemical industry before the Great Depression, Hopewell was ahead of its time relative to many areas of South. With World War II, the chemical industry expanded more rapidly in the region. The South had supplanted New England as the textile center for the United States. This in turn aided the rise of synthetic fibers, which begin life as chemicals. Fibers are processed in textile mills to form yarn and finished products. The South also possessed cheap labor, as well as abundant land, water, and other raw materials of nature. Further, energy needed to fuel chemical plants came from coal as well as natural gas in the region.

Again, federal largesse made this possible, as the government mandated spreading key war industries outside the Northeast, providing money for construction and guaranteed demand for the product. The Korean War repeated these arrangements in the early 1950s. Over half the chemical industry's plant expansions came in the South in 1951, a $600 million investment deemed necessary to national defense. With this foundation and rising demands for many types of chemicals in the postwar United States and beyond, the South continued to maintain a strong hold on chemical production. For a growing number of political and business leaders in the region, chemicals, along with technology, defense, and tourism, became markers of modernity, ways to either distract or move on from the South's image as backward and racist. As Bruce Schulman observes, "development oriented politicians almost never overtly challenged white supremacy," they "simply opposed self-defeating resistance to desegregation."[37] According to a 1969 booklet by Virginia's Division of Industrial Development, there were 164 chemical plants in the state. The industry ranked first in Virginia in terms of numbers employed (about forty-four thousand), value added, and investment in new plants and equipment. The industry grew more rapidly in Virginia than the national average, ranking tenth in the nation on the eve of the Kepone contamination. The booklet also touted Virginia's "right to work" law that kept unions weak and its lower-than-national-average spending on worker compensation. Modernity, it seems, did not have to come with increases in social welfare spending. In fact, these latter attributes businesses deemed net positives as the post-World War II South moved from being the Cotton Belt to the Sun Belt.[38]

Pesticides and the American Century

Allied became involved with pesticides after World War II, and the first synthesis of chlordecone came in 1952 from Allied chemists Everett Gilbert and Silvio L. Giolito. They noted in their patent application that their new compound could combat several pests as well as fungi.[39] Marketing chlordecone as Kepone, Allied began commercial production of the pesticide in 1958. The company manufactured it in two ways. One was from 1966 to 1974 at their facilities in Delaware and in a section of the Hopewell plant they called the Semiworks. Another was using the "tolling" method, whereby another company manufactured the final product for Allied who supplied the patented formulas, instructions, and often the raw materials. Allied used tolling to make Kepone with Nease Chemical in State College, Pennsylvania, from 1959 to 1966 and Hooker Chemical in Niagara Falls,

New York, from 1965 to 1967. This was the same Hooker later involved with the Love Canal environmental disaster in Niagara Falls, which came into the national spotlight in 1978 after chemicals from a toxic landfill damaged human health and displaced families, leading to massive federal intervention. Life Science produced Kepone from 1974 until July 1975.[40]

Why such an interest in pesticides? In the United States, Michelle Mart writes, "There has been a remarkably consistent and strong embrace of synthetic pesticides."[41] One answer is that pesticides were powerful symbols of modernity, key technologies used by the United States at home and abroad during World War II. Today's synthetic pesticides are a direct legacy of government-funded research and investment during the war. Political leaders, scientists, corporate executives, and public health officials embraced the belief that technology, developed by university-trained experts working for government and corporations, could eradicate humanity's age-old problems of famine, disease, and poverty. They were not entirely mistaken. Clearly science did bring improved health and greater agricultural production for millions. The most important example of this reliance on chemicals was DDT. Its success in World War II and after seemed to show the greatness of synthetic chemicals, conquering malaria, and killing other pests without the appearance of any residual effects. Chemicals became part of what publisher Henry Luce called the American Century—the premise he put forward in early 1941 that the United States should abandon thoughts of isolationism, enter World War II, and use the opportunity to transform the world. The reliance on pesticides took on greater urgency and importance with the Cold War, as the links between scientists, universities, corporations, and the military solidified. Chemical insecticides came to be central to these plans, particularly for their role in maintaining higher levels of food production. As Edmund Russell points out, high agricultural output "advanced defense goals" by feeding soldiers, workers, and allies; it helped stem communism by stabilizing nations susceptible to revolution during times of chaos.[42]

In a postwar American Century, amid U.S. global dominance and the Cold War, the chemical industry surged forward. "Industry's industry" provided products and connections to corporations in many industrial sectors.[43] When it came to agriculture, companies produced new pesticides and herbicides. Amid the Cold War, political leaders often framed threats to agriculture as threats to national security. Often the federal government's role was simply to approve products for rapid deployment.[44] Allied was fully invested in this national project and its ideals. Company president Fred J. Emmerich sounded upbeat in a 1956 speech outlining Allied's history and the chemical industry to the Newcomen Society of the United States, an organization of prominent business, government, and academic leaders dedicated to increasing the "knowledge and appreciation of our free enterprise system."⁴⁵ Effusive in his praise of the organic chemical industry, and with an eye toward Cold War battles for global influence, Emmerich exclaimed: "What undreamed Far Horizons of Progress have been reached—during these one hundred years! Man has triumphed through God-given vision, resourcefulness, and effort." He continued: "Out of it all emerge new wonders in the ever-amazing onward March of Civilization!" Emmerich especially extolled the benefits of the chemical industry's contributions to agriculture. He asserted, "Through better fertilizers, better feeds, and control of insects and other pests, chemistry has helped increase productivity from 25 to 100 percent, in the course of the past 20 years, depending upon the type of crop, or of poultry, or of livestock." The chemical industry was a "friend and indispensable servant to all people."⁴⁶

Others reinforced the message during the 1960s. In a 1960 publication, "Pesticides and Public Policy," the National Agricultural Chemicals Association (NACA) sought to assuage the public's growing anxiety over the effects of chemicals on human health. NACA acknowledged that it was an age of "exceptionally rapid change" and that it was not surprising that some wanted to "reverse the march of progress" out of fear. But it was a lack of dependable knowledge that added to that fear. The public, the group asserted, didn't understand how important chemicals were to the abundance of foods. Confidently they argued that keeping up with the growing world and U.S. population demanded "maximum use of chemicals in agriculture."⁴⁷

Yet there is another side to chemicals. As historian Pete Daniel notes, scientists "emerged from World War II with enhanced prestige and power, and their research and development generated seductive products that advertisers easily transformed into dazzling symbols of a better life." With the success of DDT, other synthetics quickly followed. Pesticide users trusted the advertisements that promised "insect-free homes, lawns, gardens, and fields," and they trusted federal oversight. "Conventional wisdom held that pesticides were carefully tested and, while strong enough to kill insects, were weak enough to spare fish, wildlife, and humans." But chemicals reached the "market with hardly any testing for short-term (acute) or long-term (chronic) health effects." In the federal government, the Agricultural Research Service (ARS) served as the center for pesticide approval until the EPA took over after 1970s. In that time, the ARS's "eradication schemes, oversight, and labeling reflected the chemical industry's agenda."⁴⁸ Since the Agriculture Department's primary interest lay in protecting the agricultural industry, there was clearly a conflict of interest in both regulating and promoting the same industry.⁴⁹ In Congress, those siding with industry and the USDA gained power over budgets and regulations dealing with agriculture. Chemicals went hand in hand with the increasing size and mechanization of farming. As agriculture became more capital-intensive, the size of farms grew; hand labor gave way to spraying of pesticides and herbicides, and increasingly sophisticated equipment was used to harvest crops. The publication and popularity of Rachel Carson's 1962 best-seller *Silent Spring* called into question this reliance on chemicals. While the focus of Carson's book was on DDT (which the United States banned in 1972), her work shed light on the many unintended consequences of biocides generally. As she and others began to document, acute and chronic health issues emerged for humans, and other species also suffered. Commercialization of synthetic pesticides and the earth's air and wind currents spread them across the globe. Hence inside the Green Revolution of the Cold War period and the general expansion of pesticides, government and nonprofit agencies relied on corporate pesticides and fertilizers from the United States and Europe as inputs to ensure outputs of grains and other crops, particularly in so-called Third World nations. This contributed to the increase in acute and chronic health problems in developing nations and the contamination of soils, water, foods, and livestock. Ecologically, pesticide use adds to the decline of natural enemies of pests, loss of beneficial insects like honeybees, and growing resistance to pesticides.[50]

In response to the increasing scrutiny of their effects on humans and the nonhuman world, the chemical industry mounted a vigorous defense of pesticides. This aligned with the "wave of concern about population growth" that swept the nation in the late 1960s and 1970s. President Johnson "signaled its arrival" in his 1965 State of the Union Address in which he warned of an "explosion in world population" and a concomitant scarcity of world resources, says environmental historian Thomas Robertson.[51] These concerns with overpopulation would find a voice in Paul Ehrlich's *The Population Bomb*. Published in 1968, the best-selling and provocative book sounded an alarm about "the progressive deterioration of our environment" that might cause "more death and misery than any conceivable food-population gap." Robertson notes that Ehrlich "stressed shortages" but also warned of "the damage from pesticides, fertilizers, and other high-production methods" of modern agriculture.[52]

Chemical industry spokespeople expressed concern about population growth and food shortages and asserted that more pesticides were needed. In his 1967 speech "Food for All Through World-wide Effort," later published in the *National Agricultural Chemicals News and Pesticide Review*, NACA chairman Carlos Kampmeier argued that world food production was not increasing fast enough to keep up with population growth. The solution, he said, would not come easily. "This is a complex problem, [but in] any approach that is taken ... the products of our industry—agricultural chemicals—must of necessity play a key role." American agriculture played an important role in surplus food distribution and could only do so through increased production, Kampmeier said, "with a key role played by pesticides." He added that out of almost "500 individual basic pesticides," 364 were made in the United States. The "whole world is fortunate," he asserted, for the "free-enterprise, profit-incentive system" that made this possible. Kampmeier referenced then-president Lyndon Johnson's Science Advisory Committee Panel on the World Food Supply that recommended "If food production is to be doubled in the next 20 years in the developing countries... pesticide usage will have to increase almost six-fold to 700,000 metric tons per year, [requiring] almost two billion dollars of new investment in manufacturing and distribution facilities." He concluded with the observation that famine is evidence that "nature's balance is not favorable to man." "Nature seeks to destroy man, and lack of food is its super-weapon," but intelligence enables humans to cope with this "adverse environment," that includes the creation of agricultural chemicals, despite their potential for massive damage to the environment and public health.[53]

The debate over pesticides came as chemical trade expanded. World chemical trade grew at an annual rate of 11.5 percent from 1960 to 1981, when it reached a total of $150 billion. Chemical trade was particularly important in Europe, where in 1978 Britain exported 28 percent of production, France 33 percent, West Germany 41 percent, Belgium 70 percent, Switzerland 78 percent, and the Netherlands 88 percent. The United States produced chemicals valued at $2.6 billion per year by 1980, exporting some 40 percent of that total.[54] As Kepone production was underway in Hopewell, the chemical industry overall raised prices 64 percent in 1974 and 22 percent in 1975, leading to huge profits during the 1975–76 recession.

Colorado Potato Beetle

Within this global expansion of pesticides, Kepone came to be used against two major insect pests: the Colorado potato beetle and the banana root borer or weevil. The potato beetle (*Leptinotarsa decemlineata*) played a major role in creating the modern chemical insecticide industry. Indeed, since it first began attacking potato crops in the United States in the nineteenth century, the insect has proven a resilient foe. As one scientist noted in the late twentieth century, "[The] colorado potato beetle provides one of the most dramatic examples of the outcome of shortsighted responses to a long-term problem. Due to a lack of plant resistance and biological and cultural controls, we have relied exclusively on chemicals to control this pest for over a century. It has responded by developing insecticide resistance, and it continues to be a devastating pest [throughout the world]."[55]

Despite its name, the Colorado potato beetle is native to the central highlands of Mexico, where it fed mainly on its fellow native plant, the buffalo bur (*Solanum rostratum*). Colorado became associated with the beetle first in 1865 by Benjamin Dann Walsh, pioneering entomologist who noted in the first volume of the *Prac-

The Colorado potato beetle pictured here is perhaps the best known and most widespread of the insects that threaten potato crops around the world. The beetle has proven resistant to pesticides since farmers in the 1860s first sprayed them with a paint color, Paris green, that contained copper arsenate. Kepone was but one of the many chemicals used against the insect.

tical Entomologist that his colleagues had reported seeing large numbers of the insects feeding on the buffalo bur in Colorado Territory. This convinced him—incorrectly—that it was native to the area.[56] As white settlement came to the West, so too did new forms of agriculture and food—including potatoes. Compared to the scattered buffalo bur, a farm with rows and rows of accessible, nutritious potato plants made it an easy transition for the insect. Since both the potato and the buffalo bur are part of the potato family, the beetle simply shifted to the new species. The adult is about ⅜ of an inch long, yellowish in color, with a dark orange head and ten black stripes along its back. The beetle is a voracious consumer of foliage and developed a legendary reputation for "bet hedging" reproduction, distributing eggs across space (as the beetle walked or flew within and between fields) and across time (laying eggs within and between years).[57] As Charles Mann put it in his book *1493*: "Because growers planted just a few varieties of a single species, pests had a narrower range of natural defenses to overcome." If they could adapt to potatoes in one place, they could jump "from one identical food pool to the next—a task that was easier than ever, thanks to modern inventions like railroads, steamships and refrigeration."[58] It is not clear exactly when the Colorado potato beetle first met the potato plant, but the first recorded infestation of potato plants came in Nebraska in 1859. Two years later, as the nation lurched into the Civil War, farmer Thomas Murphy of Atchison, Kansas, said that the beetles were so numerous "that they would almost cover the whole potato-vine, eating up everything green upon it." J. Egerton of Gravity, Iowa, wrote in the *Prairie Farmer* in August 1861 that they "devoured" the vines as fast as they sprung up.[59]

By the time the Civil War was over, the beetle had been ravaging potato crops in Colorado, Nebraska, and then Iowa, Kansas, Wisconsin, and Illinois. It reached the Atlantic Seaboard by 1874. In 1876, as the United States celebrated its centennial, demand for information and calls for eradication led entomologist Charles V. Riley to write what became a widely cited summation of the beetle and its movements. The book was laced with fear and references to war—themes carried forward in much scientific and popular writing on insects. The beetle, he noted, had "invaded Iowa" in 1861 and then crossed the Mississippi, in 1864 "invading Illinois," where it "occupied" and "possessed" the land. It swarmed through St. Louis in 1871, as "advanced guards" made their way into Pennsylvania, western New York, and southern Canada. Only a year after the beetle reached the Atlantic Seaboard, it landed in Europe, finding the potato crops there just as appetizing as those in the United States. Reports of the beetle came from England in 1875, Germany in 1877, and Poland in 1878. These initial beetle populations were contained, and it seemed victory had been achieved. That is, until two world wars, when the mass movement of humans, materiel, animals, and food allowed the beetle to gain a foothold it has yet to relinquish. The beetle established itself in France by the early 1920s and was common in Germany before World War II. An outbreak in East Germany in 1950 prompted a massive propaganda campaign from the East German government, claiming that the United States had dropped the beetle (dubbed *Amikafer*— Yankee beetles) from airplanes as a form of biological warfare.[60] The campaign showed that the beetle presented a crisis, given the importance of potatoes to feeding the populations of many in postwar Europe. Furthermore, unlike Napoleon or the Nazis, the beetle was able to conquer the Soviet Union by the late 1950s.

Farmers tried many ways to control or eradicate the Colorado potato beetle but to no avail. Many removed them by hand—a task often given to children— or brushed them off with brooms; some invented machines to do the work, including H. Bowen of Sheridan, Illinois, who patented a "Potato-Beetle Catcher." Others tried to encourage spiders, birds, or other insects to attack the beetle. But, as Riley noted, it became evident that "neither handpicking nor more wholesale slaughter by means of mechanical contrivances would enable the potato-grower to cope successfully with his enemy."[61] Riley himself began experimenting with different poisons, as did many farmers. They either had no effect or killed both beetle and vine. The first effective chemical application was Paris green, an arsenic compound that contains copper, which first rose to popularity as a pigment in dyes and paints in the nineteenth century. There is an oft-repeated story, likely apocryphal, that around 1867 or 1868 a frustrated farmer threw some leftover green paint on his potato plants and thus killed the beetles. The pigment in the paint was Paris green. Certainly, from about that year forward, Paris green came into use as the potato beetle made its way through the Midwest.[62]

With these beginnings, the Colorado potato beetle served as a catalyst for the modern insecticide industry and its toxic treadmill. It took about thirty years for the pattern to emerge fully in the United States. After initial success, the beetle developed resistance to Paris green, and calls from farmers and agriculturalists for more effective chemicals led the growing insecticide industry to develop new chemical applications using arsenic, including lead arsenate and calcium arsenate. When the beetle made its way to Britain, agriculturalists there adopted a similar set of technocratic solutions and deployed an array of poisons, which spread through the British Empire. Continental Europe responded to the beetle along similar lines.[63] Like the arrival of Paris green before it, the use of DDT as an insecticide promised a permanent solution to controlling the beetle. In 1939, on the eve of World War II, the Colorado potato beetle threatened the crop in Switzerland. The company that first manufactured DDT, J. R. Geigy, made available to Swiss entomologists a sample of the chemical to test against the beetle. Its success confirmed company testing, and by the mid-1940s DDT was the pesticide of choice for potato farmers. DDT had already been deployed against the mosquito to combat malaria, and the chemical became the miracle invention for a time. Companies developed other chlorinated hydrocarbons to compete with DDT in the battle against the beetle; these included Aldrin, Dieldrin, heptachlor, and then Kepone.

Banana Weevil

Kepone's second global target was *Cosmopolites sordidus*, the banana root borer or banana weevil. German entomologist Ernst Friedrich Germar in his 1824 reference work *Coleopterorum Species* first described the weevil for science from samples collected in Indonesia, then part of the Dutch colonial empire, where it originated, having evolved with the native ancestor of the modern banana. The adult weevil is dark brown to black and about ½ inch long. The adults are active at night and feed on dead banana plants, cuttings, and other decaying plant material near the base of banana plants. Adults can live up to two years and go without food for months. They walk slowly, can fly short distances, and feign death when disturbed. It is the larvae that damage banana plants by boring tunnels through the corm as they feed. This weakens the plant, making it susceptible to infections, lower fruit productivity, reduced growth, and to toppling over in strong winds.[64]

Since weevils do not move far on their own, they spread through transportation of infected planting material. From its native areas in Indonesia and Southeast Asia, the weevil spread with banana cultivation, first to Africa, then the Middle East, and eventually the Americas. Until about 1850, banana consumption remained largely confined to the tropics. Enslaved people in Brazil and the Carib-

Perhaps the most dangerous pest for banana crops worldwide is the banana root borer or banana weevil. The insect played a significant part in the development and use of pesticides, including Kepone.

bean regularly grew them for food, being well-suited to their needs since they required little labor to cultivate. After emancipation in the mid-nineteenth century, formerly enslaved people would be among the first to sell bananas to "itinerant ship captains" from North America, helping set the stage for exportation of the fruit worldwide.[65]

As it was with potatoes, with so many uniform plants in close quarters, banana plantations served as ideal hosts for plant pathogens. Yet unlike potatoes, which could serve as an important foodstuff for local populations, bananas were grown for export starting in the late nineteenth century. After first trying to move plantations to new land to avoid pathogens, banana growers settled for chemical applications. A major innovator was the United Fruit Company, which developed a spraying system using Bordeaux, a mixture of copper sulfate, lime, and water on their crops.[66] United Fruit and other major corporations also set precedents in influencing the political and economic situations in nations where they owned land and grew bananas for export. As applied to bananas, Bordeaux "set an important precedent for the large-scale use of chemical fungicides in tropical agriculture."[67] Once synthetic chemicals gained wider usage after World War II, growers deployed various compounds including Aldrin and Dieldrin. By the mid-1960s, the weevil developed resistance to these too.

After World War II, multinational banana growers shifted from the Gros Michel banana to the Giant Cavendish, a variety resistant to Panama disease. The Cavendish has a more delicate peel, so bananas were cut and packed closer to where they were produced and shipped in cardboard boxes for greater protection. Multinational growers also shifted production to Ecuador, Colombia, and Venezuela. Consumption in the United States changed as supermarkets now demanded branded and "consumer-sized" bunches from these production zones.[68] Production expanded in the Caribbean as well, especially in the French islands of Martinique and Guadeloupe and the British Lesser Antilles. Bananas were the leading export among these islands in the eastern Caribbean by the 1960s, referred to among the British as "green gold." They sold largely to protected markets in Britain and France, which allowed these industries to succeed.[69] Pesticides, including Kepone, followed.

Selling Kepone

Allied looked to market its new chemical Kepone as part of this domestic and global expansion of pesticides and commercialization of agriculture. In preparation for its possible use as a pesticide in the United States, in the 1960s the company submitted Kepone to numerous tests that revealed troubling concerns. One internal Allied letter from January 1962 that was later made public reveals that James B. DeWitt, chief of the Chemical Control Section of the U.S. Fish and Wildlife Service's research center in Maryland, informed Allied that tests done at Ohio State University showed Kepone to have the same negative effects on the human reproductive system as it did on quail and other birds exposed to the pesticide.[70] For further tests, Allied relied on Paul Larson at the Medical College of Virginia. Allied acknowledged in a 1963 memo, "The investigation of this compound reveals it to be a very toxic material if chronically ingested, with possible malignant effects." Rats tested developed tremors and had reduced life spans, and males showed diminution of the testicles. Beginning in 1971, the National Cancer Institute began testing Kepone on rats and mice as a possible carcinogen. Results showed that Kepone caused hepatocellular carcinoma in both species along with the other symptoms. However, this study did not come to public light until early 1976, only after publicity surrounding Kepone.[71]

These tests led Allied executives to acknowledge that it was unlikely that the Food and Drug Administration (FDA) would approve Kepone without reproduction studies. "[Such tests] would require nearly two years, and probably would not consider residue tolerances until the question of possible malignancy is unequivocally determined." In a company memo Allied admitted, "[The likelihood that those studies] would yield a desirable result is not good." The memo also noted the possibility of using Kepone as bait on croplands, but that would have required even more testing to show the FDA that such a use would not lead to "crop residues." Allied also listed Kepone as only an ingredient for other pesticides, not as a pesticide itself, to make it easier to earn government approval (and be less visible to the public eye). At this point, it seems Allied decided to limit the domestic use of Kepone in ant and roach traps. It had been used against fire ants in the South until Dewitt's letter of 1962.[72]

But the global market beckoned. Allied shipped Kepone globally from its Race Street facility in Baltimore, Maryland. Records from the legal cases that emerged from Kepone indicate that distributors were in Ecuador, Jamaica, New Zealand, Philippines, Australia, Puerto Rico, Venezuela, Costa Rica, Cameroon, and France.[73] Like other chemical companies, Allied appears to have established connections with scientists, research stations, and corporations engaged in the development, testing, and use of pesticides in global agriculture. In attacking the potato beetle, Allied contracted directly with the German chemical company Spiess & Sohn for Kepone use in Europe and the Soviet Union. Spiess & Sohn became a major producer and distributor of agricultural chemicals after World War II and may have used Kepone directly, but it appears the company sought the chemical as a key ingredient in making its own pesticide Kelevan, marketed as Despirol.[74] A report from the French Plant Protection Service in 1973 noted that the pesticide was tested in 1965, entered the German market in 1967, and gained some use in Spain, Austria, and Yugoslavia by 1970.[75]

Importation to France for use in the Caribbean came next. In the 1960s it became apparent to planters in Guadeloupe and Martinique that the banana weevil was developing resistance to two main pesticides, hexachlorocyclohexane (HCH) and Dieldrin. Two items came to their attention, leading them to use Kepone. First, the United Fruit Company began using the compound on its Central American banana plantations in 1965. This lasted until 1973, when the company "drastically lowered their use of insecticides" in favor of greater biological controls against the weevil.[76] Meanwhile, further tests of Kepone in the 1960s came in Cameroon under the auspices of the French research institute IFAC (French Overseas Fruit Research Institute) and the Cameroon Development Corporation. The results showed the pesticide to be highly effective.

With knowledge of these results, SOPHA, a banana company based in Martinique, made the initial request in 1968 to use Kepone. Responsibility for approving any chemical fell to the Toxics Commission of the French Ministry of Agriculture. The following year, the commission rejected its use, citing the test results from Allied that showed accumulation of Kepone in the liver and its possible risks to contaminating the environment. In 1971 France banned the use of HCH, Dieldrin, and Aldrin, but the islands received a "special exemption" for their continued use since there was as yet no alternative pesticide.[77] The commission then downgraded the toxicity level of Kepone, and it received approval from the Ministry of Agriculture in 1972 (under the direction of future French president Jacques

Chirac). That year a French company, Vincent de Lagarrigue, agreed to import Kepone for use in the French West Indies, establishing the connection between Allied, Hopewell, Kepone, and the islands.[78]

When William Moore and Virgil Hundtofte agreed in 1973 to toll Kepone for Allied, global markets such as the French Caribbean were there for the taking. They were ready to supply a weapon to support those like Carlos Kampmeier who viewed nature and insects as the enemies of humans. His views recall what Joshua Buhs wrote in his work on the fire ant: "Pests are not born; they are made."[79] It is not only the real damage insects cause to crops that is important but also the associated cultural expectations regarding insects. Historian James Giesen makes this point in his work on the boll weevil in the American South. The history of the weevil's spread across the region "[is as much about] fear and expectations as it is about reduced crop yields." The insect's "greatest consequence was not the many strands of cotton it devoured, but rather the great explanatory power that people found in the weevil."[80] These insights bear directly on the Kepone story. The two insects central to Kepone's use, the Colorado potato beetle and the banana root borer, coevolved with the two crops they consumed, becoming pests of growing significance to the human reliance on potatoes and bananas. The destruction caused by Colorado potato beetle and the banana weevil was real, but so too were the ways in which those insects reinforced ideas in the decades following World War II about geopolitics and the growing reliance on techno-scientific solutions that reinforced powerful networks of experts, policy makers, and corporate leaders. Pesticides embodied these cultural aspects regarding insects, combining with insect and plant biology and the powerful synergies of science, government, and business.

Kepone brought Hopewell, a city with rich connections to agriculture and chemicals, into this pesticide history. When it entered the market, Kepone filled an important niche in the global battle against insects. Yet by the early 1970s Kepone remained largely unknown to the general public and the wider governmental and scientific community. Likewise, while boosters claimed it to be the Chemical Capital of the South, Hopewell also avoided the national and international spotlight. But this invisibility changed dramatically in 1975. Afterward, Kepone and Hopewell would no longer be in the shadows.

CHAPTER 3

The Kepone Shakes and a Poisoned River

A front-page story in the May 12, 1975, *Hopewell News* came with the headline: "Bugless sludge put in special pit." It was accompanied by a picture of Hopewell sewage treatment supervisors watching a pump fill a pit with sludge laced with Kepone. The caption offered a hopeful statement: "Resolution of the Kepone problem nears its end." According to the story, the year before, Life Science Products (LSP) had discharged enough Kepone into the local sewers that the chemical began killing the beneficial "bugs" or bacteria that digested the city's solid waste before it could be treated and discharged. The result was untreated municipal waste. As the article explained, the city was now emptying its two digesters, filling them with new bacteria, and getting both back in operation. Public Works director Fred Hughes Jr. was confident the system would be working again within two months.[1] His optimism would prove misplaced.

The bugless sludge wasn't the first indication something was amiss with Life Science, nor would it be the last. Faulty seals at LSP exposed people and the environment to sulfur trioxide. Kepone dust often blew across Randolph Road, and LSP workers already exhibited the "Kepone shakes." Under the direction of company leaders, LSP workers were busy dumping Kepone waste into sewer drains, open pits, and the landfill in Hopewell. Then, in June 1975, Dale Gilbert, sick with Kepone poisoning, would visit Dr. Yinan Chou at his wife's behest, setting into motion events that brought Kepone into public scrutiny. Chou, a native of Taiwan and recent hire at John Randolph Medical Center in Hopewell, sent blood and urine samples to the Centers for Disease Control (CDC). The CDC then contacted Dr. Robert Jackson, a young, outspoken physician at the Virginia Department of Health. Jackson promptly investigated the situation and discovered several workers ill with Kepone poisoning and found Kepone dust contaminating the Life Science facility.

The actions taken by the Gilberts, Chou, and Jackson—lay people and experts—brought to light the immediate and longer-term concerns over Kepone. Before this, other local, state, and federal authorities had already been aware of either all or part of what was happening inside the facility yet for different reasons had failed to shut things down. Examining the initial discovery of Kepone poisoning and contamination reveals the inconsistencies and weaknesses in the environmental management state, showing the tenuousness of assurances regarding control over chemical environments. The many scientists, corporate officials, and government authorities constructing the chemical age relied on a discourse equating chemicals with progress and critical for victory over such things as disease, pests, and communism. They insisted that chemicals were safe for people and the environment. Moreover, as Christopher Sellers reminds us, in areas of workplace and industrial hygiene, experts, government, and corporate leaders all assured their workers and those listening and watching that "industrial threats to the human body were coming under control."[2] Kepone wasn't the first or last chemical for which such assurances did not hold true. Kepone illustrates what Sonya Boudia has said about chemical residues: "[They are] transgressive. They disobey boundaries, appear where they shouldn't appear, alter environments, and enter communities and bodies without permission."[3] Chemicals ignore boundaries and laws, moving through the air, soil, and water, penetrating human and nonhuman bodies. The initial discovery of Kepone showed the motility of the chemical between two usually separate areas of safety and regulation: workers' bodies and the environment.

It is also important to note that chemicals are transgressive not just by themselves, drifting with wind or water currents, but also in collaboration with humans responsible for their creation, manufacture, use, regulation, and disposal. Continued investigations revealed these collaborations, showing deeper aspects of the Kepone tragedy.

From Allied to Life Science

Two men at the heart of the Kepone story are the founders of Life Science, William Moore and Virgil Hundtofte. William Moore died in Hopewell in 2004. A native of North Carolina, Moore was twenty-one years old when he graduated with a bachelor's degree in chemical engineering from North Carolina State in 1945 and then went to work on the Manhattan Project at Oak Ridge, Tennessee. After the war, he went to Allied Chemical in Hopewell where he spent his career, earning a master of science degree from the University of Richmond in 1956 along the way. He retired in 1973 as director of research for the Agricultural Division in Hopewell. He developed a number of patents that, according to his obituary, "enhanced the chemical industry during his tenure with Allied Chemical Corporation and other patents that improved the development of products for the environment, humankind, and animals."[4] His wife, Vera, is a descendant of William Claiborne, who emigrated from England to Virginia in 1621 and served in a number of capacities in the colony's affairs.

Background on Virgil Hundtofte is harder to locate. From available sources, it seems he was born in 1937 in Sidney, Montana, to a large farming family. He earned a bachelor of arts degree in chemistry from Pacific Lutheran University and a bachelor of science in chemical engineering from the University of Washington in Seattle. He came to Allied in 1965. He was the plant manager for the Agricultural Division in Hopewell, with about four hundred employees. This included the Semiworks, where Kepone was made from time to time starting in 1966. After the Kepone incident, he moved to New Mexico, then Scotland, and then Kuwait, where he became involved with TIL, a company he cofounded that does consulting on career development and training for employees.[5]

At different points, both Moore and Hundtofte had responsibility for Kepone production at Allied. As Allied Agricultural Division president G. C. Matthiesen stated in the 1976 Senate hearings on Kepone contamination (before a Subcommittee of the Committee on Agriculture and Forestry), international demand for Kepone increased in the late 1960s. Allied brought production in-house from former tolling operations at Nease and Hooker Chemical "to occupy open time" at Hopewell's Semiworks, a pilot plant at the sprawling facility used "for the development of processes to be used in larger scale commercial production of various chemicals."[6][7] The Semiworks produced Kepone from 1966 to early 1974 along with other products, especially two used for coating electrical wiring, THEIC (Tris [2-hydroxyethyl] isocyanurate), and TAIC (triallyl isocyanurate). Initially Kepone fell under the Agricultural Division's research area, where William Moore held ultimate responsibility from 1966 to 1969. Allied then moved Kepone to the manufacturing area within agriculture, during which time both C. L. Jones (who would later become Hopewell's liaison engineer for the sewage treatment plant) and then Hundtofte supervised Kepone production. Then in 1972 Allied executives decided to shift the Agricultural Division to Houston and place Kepone production in Hopewell under the authority of the Plastics Division. At this point, Hundtofte resigned rather than move to Houston. Moore retired from Allied in 1973, although he maintained a consulting relationship with Allied until he started Life Science.[8]

With Allied's Plastics Division now controlling the Semiworks, those executives had to decide whether to continue to make Kepone or to make THEIC. According to G. C. Matthiesen, THEIC became the priority, and Kepone would either need a new tolling operation or Allied would cease to make it. All the tragedy that would come from Kepone happened over what was a small part of Allied's global presence. Profits from Kepone never exceeded $200,000 a year—a drop in the bucket of Allied's $3 billion in annual sales. Yet there was demand from Europe as well as the Caribbean and elsewhere for Kepone, with major sales going to the companies Spiess & Sohn in Germany and Vincent de Lagarrigue in

France, so Allied decided to outsource and stay in the Kepone business.[9] Tolling operations were common for Kepone, as they were for other chemicals. Among other benefits, the liability shifted to the company doing the processing. As Matthiesen stated in the case of Life Science, that company was responsible "for the risk and liability arising out of their operations."[10] After deciding to toll, Allied executives looked for a company to take over Kepone production and contacted Nease Chemical, Velsicol, and Hooker. They also met with William Moore about the possibility of making Kepone. Moore contacted Hundtofte, and they created Life Science Products in hopes of winning the bid. Nease and Velsicol declined the opportunity. Life Science obtained the contract with a much lower bid than Hooker—54 cents per pound to Hooker's $3 per pound.[11]

The contract between LSP and Allied became a key piece of evidence in the later legal proceedings. At its own expense, Allied provided all the raw materials for Kepone production, and LSP produced Kepone solely for Allied. Allied set monthly production goals and set the specification requirements for Kepone's purity. Allied required 650,000 pounds minimum in the first year and established 3,000 pounds per day as the maximum in the first year. The Kepone would be packed in Allied containers and moved to Allied's Hopewell facility on Allied trucks. Allied agreed to reimburse LSP for all taxes, whether federal, state, or local, on the "conversion, production, warehousing, transportation, delivery, or use" of raw materials supplied by Allied. Also, Allied agreed to pay higher rates if LSP needed additional equipment or capital for environmental controls. If LSP were to close for any reason in the first year for pollution violations, Allied could purchase LSP's assets for $25,000.[12] The contract deepened the ties between Moore, Hundtofte, Kepone, and Allied, and those connections would grow as the investigations and problems involving LSP's pollution mounted.

After obtaining the contract with Allied, Moore and Hundtofte had to next decide where to build their manufacturing facility, obtain the equipment, and hire workers. They arranged all this quickly, between November 1973, when the contract was signed, and February 1974, when production began. For their facility, they chose a site not far from the massive Allied plant on Randolph Road that included a vacant gas station and a service building with an open lot next to it. The site was, according to Moore, "the only one that we could find available in the timeframe and cost frame that we were working in."[13] They renovated the existing buildings into an office and a production area. They added two outbuildings, described as "shed like enclosures," "grouped around an open paved area" known as "the pad," where intermediate steps in Kepone production were located.[14] Deputy health commissioner and later head of Virginia's Kepone task force James Kenley described the setup as "several corrugated metal over steel frame structures, and tanks located in open areas." Moore later stated that he and Hundtofte pur-

Life Science Products building. The owners of Life Science Products converted a gas station and adjoining land into their plant to manufacture Kepone. This image shows the facility just before the main office building was cleaned, wastewater in the holding ponds emptied into rail cars, and the remaining cement for the ponds and the processing equipment dismantled and buried in lined pits in Hopewell. The site is currently used for an automotive repair shop. AP photo by Mary Ann Carter

chased equipment "from all over the country" and that it was "as good as could be bought" and not second-rate. But Life Science worker Thurman Dykes considered the overall setup a "fast put together job" to "get production out real fast."[15]

Moore and Hundtofte saw Kepone as only a first step in growing their new company. Moore focused on improving the process of making Kepone and doing research for new ventures for Life Science, while Hundtofte took over the main responsibilities for the day-to-day running of the plant and hiring workers. Initially he brought in several workers from Allied on a part-time basis to start the company, both operators and supervisors, as well as engineers. It's not clear how the shift workers were hired or how they learned about jobs at Life Science. It may have been personal connections and word of mouth. Several had worked at Allied before, and it seems there was a layoff at one sector of Allied right around the time Life Science began hiring. Chemical processing is generally not labor-intensive. With a fluctuating workforce of about thirty to thirty-five at any one point, Life Science operated around the clock to meet demand. The small facility produced some 840,000 pounds from March to December 1974 and some 846,000 from January through July 1975. At the LSP facility, Hundtofte worked directly with the operators from February 1974 through July 1974. Then he hired a superintendent, and Hundtofte began spending more time with Moore on their new ventures but still came to the Kepone plant. In April 1975 he hired a plant manager so he could focus most of his time on the new ventures for LSP with Moore. Delbert White began working on a part-time basis as plant superintendent and then took that job full-time until April 1975. From April until the plant closed in July, H. D. Howard held the job. Another key person with Life Science was Richard Ogden, a chemical engineer who had set up Kepone production for Allied in the Semiworks.[16]

The Kepone Shakes

It was the end of June 1975. Radios in Hopewell might have been playing "Love Will Keep Us Together" by Captain and Tennille, "The Hustle" by Van McCoy, or "Rhinestone Cowboy" by Glen Campbell. Many residents were likely headed to see *Jaws*, which opened nationwide June 20. We don't know if Dale Gilbert listened to those songs or saw *Jaws*, but we do know that in June, Gilbert, an operations supervisor for Life Science, became too ill to work. At thirty-four, he was a physically strong, sturdy man. He and his wife Jan had three children, ages thirteen, twelve, and eleven. Raised in the mountains of Southwest Virginia, Gilbert had worked in Hopewell since the late 1960s. He had been employed as a supervisor at a warehouse for tobacco makers Brown and Williamson in nearby Petersburg when his neighbor, Delbert White, asked if Gilbert would like to work for Life Science. White, a former marine, had also worked at Allied and became a plant supervisor at Life Science in the summer of 1974. "At the time," Gilbert said, "I thought it was a good move." He had been making $250 per week, and the new job at LSP brought in $278.50 per week. Although he worked long hours, according to his wife he was a man who had "always been able to handle" what came his way.[17]

Gilbert began at Life Science on a part-time basis in August 1974, then became full-time in January 1975.[18] He noticed his hand and body tremors about two weeks later, by March he began having severe chest pains, and by June he was too ill to work. He also had "twitching eyes and weight loss." Gilbert had developed the characteristic symptoms that his coworkers had already named the "Kepone shakes." Despite the installation of a bagging operation designed to capture Kepone dust during production, it coated the production area of the plant. When it got wet, it became a thick slurry. LSP workers inhaled the dust, swallowed it, and absorbed it through their eyes and skin. According to Gilbert, the plant "was like a duststorm most of the time," with the Kepone powder covering everything.[19] Some called it the "flour factory." Other workers reported a similar situation. Tom Fitzgerald, who had served in Vietnam, worked at a dryer and packed the Kepone powder into barrels. The pace left little time for breaks, so he remained at his station eating his lunch "between clouds of dust." Decades later there were certain memories that were especially vivid: "The way the place smelled. It was a dusty—I don't really know how to describe it. It smelled like flour that had gotten wet and gone bad or an old jug of Elmer's glue, pretty much the same thing, I guess, just a rawness to it, a sense of decay."[20] Thurman Dykes, like Fitzgerald, operated a dryer in what they called the bag house. Dust "settled on just about everything," he said. Kepone would "stick to your body." Management, Dykes said, never told him about the dangers of Kepone, only to be careful handling the acids that went into making Kepone. As Fitzgerald said: "Management admitted that the reason that everybody was shaking was because of Kepone," but they assured him it was not anything to worry about. "Del White had the shakes himself," Fitzgerald said, but White didn't think anything of it.[21]

Dust also blew away from the plant before Life Science installed dust collectors, but collectors could break down or become clogged, and then dust escaped again into the air. Kit Weigel worked as a reporter for the *Hopewell News* in 1975 and later became its editor. The newspaper's building was diagonally across the street from Life Science. "There'd be all this particulate matter falling on our cars," Weigel told me, "and I'd go out at night and dust it off the windshield so I could drive home."[22] Dust settled on businesses and homes, on an ice-making facility across the street from LSP, and in soil in Hopewell. Later investigations showed Kepone dust as far as forty miles away. But people did not think much of it at the time, since releases like that came from other, larger plants at various times too. Reverend Curtis Harris, a noted civil rights leader who was a confidant of Martin Luther King Jr., had his home and his church not far from Life Science. He had to close the church windows to keep the dust out. "We assumed this would last for a short period and blow over," he commented. "If you live in an industrial area, you tend to pay less attention."[23]

By comparison, it seems from available accounts that the same process of making Kepone at Allied did not result in massive contamination on the shop floor or workers suffering with shakes and tremors. Allied chemical engineer Jameil Ameen supervised Kepone production at Allied from 1966 to 1969, and his boss was William Moore. Ameen stated in the 1976 Senate hearings on Kepone that his "office was about 100 feet from the area of the plant where Kepone was produced" and that it was kept "neat and clean." Further, he claimed, "[Moore] insisted that we do everything by the book," and there was "careful attention to detail." Employees, it seems, wore proper safety gear, showered after their shift, and cleaned up any spills quickly. If there were serious problems with workers, they did not come to public view.[24]

After his frustration with being prescribed tranquilizers by one of the doctors regularly used by LSP and Allied, Gilbert sought out cardiologist Yinan Chou. Chou remembers Gilbert coming to him for chest pains and noticed "something odd": tremors and rapid eye movements that suggested something else besides his heart problems. "I was puzzled," Chou recalled to me decades later. Unlike others, Chou began asking questions about Gilbert's job. "I asked him 'What kind of job do you do?'" He said he worked at Life Science where they made a pesticide. Chou sent Gilbert to the local hospital, John Randolph, for more testing. Chou recalls that about "a day or two later, Mr. Gilbert handed me a small piece of white label with 'Kepone' on it."[25] Like many others, Chou had never heard of the substance. He did research in medical journals and "pharmacology or toxicology books" for information and found that Kepone was related to DDT, which could cause tremors, weight loss, and anxiousness in rats. Chou suspected Gilbert had Kepone poisoning. Neither the local hospital nor the Medical College of Virginia (MCV) could test for Kepone in the body. "The lab technicians at the [John Randolph Hospital] lab were fantastic and were enormously helpful," Chou remembers. They called commercial and state labs in Virginia, but none were equipped to perform the test. Chou then called the Center for Disease Control (CDC) in Atlanta, and he sent samples of Gilbert's blood and urine there. "It was fortunate that I sent the blood as well," Chou told me. The CDC at first only recommended urine, but, as it turned out, Kepone doesn't present itself in urine.[26] A newspaper article from January 1976 noted that Chou was initially reluctant to have his name used in the media for fear he might get in trouble with other physicians.[27]

When the results came back at the CDC office in Atlanta on July 18, two doctors there made phone calls. Dr. Renate Kimbrough contacted Dr. Chou at his office. Chou told me, "[Dr. Kimbrough] was almost in panic and said 'the blood test showed your patient is intoxicated with Kepone!'" Chou then called Gilbert. He said, "Mr. Gilbert, you are intoxicated by Kepone! I want to refer you to MCV for treatment." And Mr. Gilbert promptly became a patient of Dr. John Taylor, a neurologist at MCV."[28] Taylor remembers: "It was obvious that he had a new and unique problem. He actually sat kind of across the room from me, I remember. He called it a Kepone shake. I said, 'Well, what do you mean by that?' He said, 'Well, everybody who works there has them.' I said, 'Hm . . .' And that kind of got my attention."[29] After examining Gilbert, Taylor called the nearby Central Virginia Poison Center and confirmed Chou's diagnosis. Taylor noted that Gilbert had "had 40 pounds of weight loss, tremors, unusual eye movements, a rapid pulse, and a tender enlarged liver." Taylor also feared Gilbert might have suffered brain damage.[30]

Meanwhile, Dr. Edward Baker of the CDC contacted Dr. Robert Jackson at the Virginia Health Department on July 18 and reported results of the tests done on Gilbert's blood—7.5 parts per million of Kepone. Originally from New York, Jackson was thirty-three and known to speak his mind. James Kenley remembers him as a "bright young fellow, one of the brightest workers I had." He was also "very much of a showman."[31] Jackson called Life Science and arranged to tour the plant and examine workers coming off their shifts on July 23. He also requested that Virginia's Bureau of Industrial Hygiene visit Life Science to take atmospheric samples. He learned from that office that their staff had finally been able to schedule a visit of their own, which was planned for later that same week. This stemmed from the May 1975 incident when a worker at the Hopewell Sewage Treatment Plant was overcome from fumes of HCP, one of Kepone's ingredients. Jackson first went to the office of three Hopewell general practitioners near the plant, he said. "The first man I saw was a 23-year-old who was so sick he was unable to stand due to unsteadiness, was suffering severe chest pains, and on physical examination had severe tremor, abnormal eye movements, was disoriented and quite ill." Jackson had him admitted to MCV under John Taylor's care. He examined nine others, six of whom had varying degrees of Kepone poisoning. He then went to the plant on Randolph Road. He toured Life Science with one of the industrial hygienists who had taken atmospheric samples, James Saunders, and Life Science plant manager H. D. Howard. Jackson, wearing "rubber galoshes, waded through puddles of Kepone-contaminated water in a makeshift factory filled with Kepone-caked machinery." On the tour, Jackson asked about protective equipment for the workers; the few workers there only wore hard hats. "I asked particularly about a respirator, because in the drying room of the plant the dust was quite thick." According to Jackson, "Howard went to a desk, lifted a mound of papers, sort of shook up the dust, and came up with three small light plastic dust protection devices, one of which had a broken strap. It was clear they were not used, and the workers confirmed that they very seldom used any kind of protection."[32] The next morning, July 24, Jackson met with James Kenley and received confirmation that, according to Virginia law, the Virginia Department of Health could order the plant to cease operations. Armed with that information, that afternoon Jackson, along with officials from the Bureau of Industrial Hygiene and the attorney general's office, met with Hundtofte and Moore who then agreed to halt further production before being ordered to do so by the Health Department.[33] Moore and Hundtofte also agreed to cooperate in cleanup of the site. It seems either Virginia authorities or the federal Occupational Safety and Health Administration (OSHA), which became involved at this point, or perhaps both, allowed Life Science to finish production of the remaining Kepone until September 1975 with workers using proper safety equipment such as respirators and under supervision of both Virginia and OSHA authorities.[34]

It was a fast response from the Virginia Department of Health, but many wondered why workers at Life Science had been exposed for so long and why no others besides Chou acted. In the U.S. Senate hearings, Alabama senator James B. Allen was exasperated over this point. "I am amazed it took 16 months with the employees reporting these symptoms for it to be established what was causing this con-

Crew cleaning up the Life Science Products plant in Hopewell, Virginia. Workers from Allied Chemical were responsible for cleaning and dismantling the Life Science facility in the wake of the Kepone disaster. The protection they are using in this image is in stark contrast to the general lack of safety equipment used during Kepone production at Life Science Products. © 1976 *Richmond-Times Dispatch*

dition." While he was not shy about making waves, at the time of the incident Robert Jackson defended the Hopewell medical community. "I would like to state that this disease is an extremely unusual one, which had not been described in the medical area very well. One would not expect the average practitioner in the area to recognize it as what it was. The symptoms could have been caused by any of a number of other causes, other than Kepone."[35] In his later interview in 2019, Chou also defended local doctors. He remembered that there "was some concern at the time" about a "conspiracy theory" or cover-up among Hopewell doctors to not report symptoms or follow up with further testing. "I'm sure none of these are true." Ethically, the doctors had upheld the "highest standard" he had ever seen, he said. He added that the poisoning was "challenging," basically "a new disease" that "was not a typical textbook case." He felt that his responsibility had been to "help the patient and referring physician. Therefore, I [had] to be much more attentive, careful, and be more vigilant. I [had] to think beyond the normal boundary."[36]

After closing the plant, Jackson focused next on the Life Science workers who had not been tested. During the first week of August 1975, he, nurses, and another doctor, Shanklin B. Cannon from the CDC, were able to contact about 110 who had worked at Life Science. By the end of the year they had reached about 149 workers. Of those, Jackson reported that 76 had some illness from Kepone, and 29 of those were hospitalized at MCV. Doctors sent blood samples of those in the hospital to the Virginia state lab for analysis. Blood from the other workers went to the CDC for testing. Given the numbers and novelty of testing for Kepone, it took months to complete all the tests. Fourteen of the men were diagnosed as sterile.

As Dr. Sidney Houff, one of the physicians attending to the Life Science workers at the Medical College of Virginia noted: "To be sterile for any man, I think, is a very emotionally devastating thing. Many of these patients are young. They have never had an opportunity to complete their families even if they started. They cannot now look forward to completing those families with any type of certainty."[37] They were also worried about the children. The Dykeses had a son in September 1975, born with a Kepone level of .3 parts per million and a liver that would not function properly. They feared long-term damage to the child.[38] Tom Fitzgerald remembered that fear. His Kepone levels were high, "ten times what was considered dangerous" he says. "I think one of the scariest diagnoses that I got... I was twenty, and they told me that by forty I would probably be riddled with cancer."[39] A team of doctors and nurses from Virginia's Health Department and the CDC also conducted tests on residents and workers from businesses in the area of Life Science up to a mile radius to determine how widespread the poisoning might be.[40]

Although many doctors, political leaders, and regulators stated they had never heard of Kepone, there was some data available, even if it wasn't generally circulated. Kepone was listed in Dreisbach's widely used *Handbook of Poisoning*.[41] Allied had conducted tests on the pesticide in the early 1960s, and those results were available at Allied and at Life Science but not publicly shared. The main conclusions from these tests were that Kepone, like many other pesticides, not only affected the central nervous system and reproduction but also caused cancer in laboratory animals and possibly in humans as well. Dr. Taylor noted, though, that there was no "human reservoir" of testing or documentation of effects to draw upon for Kepone, other than that which they were observing at that moment in Hopewell.[42]

Closing the James

As testing of the poisoned workers went forward, Dr. Jackson requested assistance in beginning the environmental sampling for Kepone. He first contacted the EPA's Health Effects Research Laboratory in Research Triangle Park, North Carolina. That office sent a team to Hopewell to assist Virginia's State Water Control Board (SWCB) as they dispatched biologists to gather sediment and fish samples around Hopewell and in the James River. Dennis Treacy and David Paylor were working at the SWCB while in college and were among those who gathered samples. Treacy is now senior counsel at Reed Smith LLP, a law firm in Richmond, where he specializes in government relations. He finished his degree at Virginia Tech in 1978 and gained his environmental law degree in 1983 from Lewis & Clark College. He served as executive vice president and chief sustainability officer of Smithfield Foods and from 1998 to 2001 as director of the Virginia Department of Environmental Quality.[43] "We really didn't know what we had [with Kepone]," he recalled. "We worked at a place called the Division of Ecological Studies [DES], and we were the firefighters. We were the biologists that actually went in the field when something went wrong, which was every day." [44] Kepone was one of many environmental crises across Virginia in the early 1970s. David Paylor, former head of Virginia's Department of Environmental Quality (DEQ), also began his work as a field biologist with the DES. He later worked with Treacy at the DEQ and led the agency from 2006 to 2022. "A lot of things went wrong. We had three hundred fish kills a year that we had to investigate. But [Kepone] was the big kahuna because it was the whole James River and most of the Chesapeake Bay. We didn't know how much was safe and what the half-life was and how quickly it was going to go away or any of those kinds of things, so those were the kinds of questions that we were being called on to answer. How much is in the fish?"[45]

Indeed, the testing facilities had to cope with samples from workers and hundreds of samples from sediments and fish, looking for a substance with which they had little or no experience. Part of the requirements meant first ensuring comparable testing at labs performing the work. The EPA worked mainly with Consolidated Labs in Richmond. Finally, the EPA published results of the environmental testing in Hopewell and the James River in mid-December 1975. They were significant and made clear the scope of the crisis and that the negligence went far beyond poisoned workers. "Kepone in the water of the James River was detected at levels between 0.1 and 4 parts per billion (ppb). Kepone was detected at levels between 0.1 and 20 ppm in fish and shellfish taken from the James River." Some of these were as far away as forty miles from Hopewell. Carp and catfish contained .01–.2 ppm of Kepone, while shad and bass contained 1–2 ppm in the filet, but the liver of bass was as high as 14 ppm. Clams and oysters contained .2 to .8 ppm. "Sediment from the Bailey's Creek area and waste water from the sewage treatment and landfill area contained Kepone at levels between 0.1 and 10 ppm. Sludge samples from the holding pond and from the landfill near the Hopewell sewage treatment plant contained 200–600 ppm of Kepone." Soil samples immediately adjacent to Life Science along Terminal Street (now Rev. C. W. Harris Street) "contained 10,000 to 20,000 ppm of Kepone." Water in Bailey Creek showed levels of 1–4 ppb. Water at the holding pond on the LSP site contained 2–3 ppm. In addition, ice from an ice plant across the street from LSP contained 0.1 to 1.0 ppm of Kepone. Luckily, drinking water in Hopewell contained no detectable levels of Kepone. Air filter samples gathered between March 1974 and April 1975 from a Virginia air monitoring sampler about two hundred yards from the Life Science plant contained between 0.2 to 50 micrograms per cubic meter of air.[46] The CDC did not release full results from the blood samples taken from residents in Hopewell until March 1976, so residents there had to wait seven months before knowing if they were in trouble. When complete, of the 216 residents tested, 176 had non-detectable levels. The remainder averaged around one thousand times lower than that of the ill workers, about 10.9 parts per billion.[47] Still, the fact that Kepone dust could affect citizens in the city was troubling.

In December 1975, there were not yet allowable "action levels" for Kepone in fish and shellfish. There was no known "safe" level for it in humans at all, so regulators were operating amid great uncertainty and pressure. The action levels would come in February 1976 and be set at .3 ppm for shellfish and .1 ppm for fish. In March 1976 the EPA set the level at .4 ppm for crabs. An action level is set for pesticides added to food outside regular agricultural practices and purposes. For pesticides added to food through use in agriculture, the EPA and FDA set a "tolerance level." Action levels were thus considered temporary and could remain until more testing and knowledge came forward, at which time a steadier tolerance level could be set. The EPA and the FDA worked together to set these levels—the EPA recommending a level of a certain pesticide or chemical, and the FDA, responsible for food, deciding whether to accept the recommendation. At the time of Kepone, the EPA determined the action level of a substance (usually in parts per million or billion) by first relying on toxicology studies to discover the smallest amount of a substance that would produce an adverse effect in test animals (usually rats or mice, sometimes dogs), and then reducing that by a safety factor of 100 to account for population that included old people, ill people, and infants, and to account for the range of susceptibility of test animals. The EPA could reduce it to 1,000 if the substance has potentially severe effects on reproduction or causes cancer.[48]

Environmental testing for toxic substances had grown out of research and publicity on the residues of DDT, which reached greater public awareness as a result of Rachel Carson's 1962 book *Silent Spring*. Virginia responded to growing public environmental concerns and new federal laws with its own emphasis on battling pollution and cleaning up waterways. Treacy and Paylor remember that historical moment in Virginia and how they both participated in the emerging environmental management state. Treacy recalls that Virginia had a "very aggressive" Water Control Board. Problems with pollution came to light in waterways like the James. "Remember, it's the early '70s, and people were interested in factories and making stuff, but it was the time when people began to notice in earnest the environmental externalization of a lot of the costs that were associated with this. That's how that was. Our work was based on science, and what made us go do our work was public opinion sometimes, but, at the end of the day, we weren't making stuff up. We weren't guessing. We were trying to do God's work here."[49] For Paylor, who was at Duke University, Carson's book was influential. "So '71 was just a few years after Rachel Carson and *Silent Spring* and all of that, and I took an ecology class and just sort of naturally fell into wanting to do environmental stuff." Recalled Treacy: "Our job was to—sounds corny or self–whatever it is—save the citizens of the commonwealth from this poison. That's what we thought we were doing."[50]

While the public gained more information on DDT and other pesticides, ignorance reigned when it came to Kepone. "It was scary," remembered Treacy. "It was different than oil, and it was different than mercury, because everybody knows what that is, and you can say, well, oil covers the duck, and the duck is dying, and that's what reporters tend to show. This one was a complete unknown. Nobody knew what it was. Nobody knew what it would do. The Rachel Carson thing was still in the back of everybody's mind. What is this?"[51] Robert Jackson noted similar feelings at the time when he testified at the 1976 Senate hearings. "We have little or no information on the long-term effect of human exposure to low levels of the material, or what level represents risk." It also turned out that the state Air Control Board did not even measure for Kepone, since it was not among those chemicals listed for measurement. When asked, Kenley, Jackson, Governor Godwin, and other regulators in Virginia and at the federal level admitted that they possessed little to no knowledge of Kepone or its effects before hearing about the poisoned Life Science workers. Even Huntofte and Moore said they lacked complete knowledge of the very thing they produced.[52]

Dr. John Finklea, director of the National Institute for Occupational Health and Safety, part of the CDC, admitted as much at the time. "There are still many unanswered questions about the fate of the workers already exposed to high levels of Kepone. At this time, we do not know to what degree their clinical symptoms are reversible. Also open to question is the possibility of delayed health risks. We have evidence that Kepone causes cancer in animals. We have as yet no evidence that it causes cancer in humans. We have evidence that occupational exposure to Kepone affects the male reproductive system. We do not yet know whether it has other adverse effects on the workers' wives and children." Finklea went on to consider Kepone in a larger context. As he noted, ignorance of the short- and long-term effects of Kepone on humans was no isolated thing. Despite the continued development and use of pesticides in the United States and worldwide, regulators, city officials, scientists, and citizens lacked knowledge about their effects on animals and human beings. "Unfortunately, the tragedy that occurred in Hopewell is probably not an isolated incident. There are approximately 100 pesticide manufacturers in the United States, about 3,000 formulating operations, and tens of thousands of applicators. Although there is ample information on the effects of pesticides on pests and some information on the acute effects on man, little is known about chronic effects of human exposure to most pesticides. Inconclusive data suggest, however, that many pesticides in addition to Kepone may be carcinogenic in man."[53]

Because of what happened in Hopewell, we now have much more information on Kepone. Allied had already conducted tests on Kepone in the early 1960s, but those were not publicly available. These showed that Kepone produced cancer in tested rats. It also impaired fertility among male and female rats. The Fish and Wildlife Service had by then showed how it altered the biochemistry of pheasants, giving male birds female sexual traits.[54] The National Cancer Institute (NCI) had begun studies of Kepone in late 1971 or early 1972. The results were complete in 1973 but not officially released until early 1976—months after Kepone contamination was well underway. The NCI results showed Kepone producing cancer in mice and rats, along with impaired fertility, tremors, and other ill effects.[55] Debate remains as to why the NCI results did not come out in 1973. Some indications are that the Allied asked the NCI not to release the positive results, and it may be that NCI was reluctant to share positive results. During the House hearings on Kepone, Sheldon Samuels, director of health, safety, and the environment from the AFL-CIO lambasted the secrecy and delay.

> I have read the three reports submitted to the Federal Government by Allied in 1960, 1961 and 1962. There was clear evidence that Kepone is a carcinogen. One of the consulting pathologists was an employee of the National Cancer Institute. Four members of the faculty of the Medical College of Virginia were also consulting scientists. Yet nothing was done by Government or management—no research, no regulation, no voluntary action—aimed at Kepone as a carcinogen as a consequence of these reports....
>
> The fact that an NCI employee had evidence of Kepone carcinogenicity in 1961 had no effect on the NCI bioassay program. Nothing was put in the open literature or available to other NCI scientists, who were unaware of the 1962 data because it had been submitted in confidence, in secret.
>
> NCI began its study of Kepone in 1971. The determinations of carcinogenicity should have been publicly released in 1973. The system of contractors, subcontractors and the reluctance of the institute to permit NCI scientists to expedite findings of any kind resulted in a release of this information only because of the glare of the Hopewell incident.[56]

Governor Godwin became aware of the situation in the summer of 1975, once the Health Department became involved. Otis Brown served as secretary of health

Members of the Kepone task force. Dr. James Kenley is standing, and to his left is Dr. Robert Jackson. Governor Godwin created the task force in December 1975 to manage the response to the Kepone disaster. Courtesy of Dr. James Kenley

Governor Mills Godwin (*center, with binder on lap*), seated with Dr. James Kenley. The date of this photograph is not known, but it captures one of the many times both men spoke or testified about the Kepone disaster from late 1975 through 1977. Courtesy of Dr. James Kenley

and human services under Mills Godwin and became the governor's liaison with the many aspects of the Kepone situation, work he continued at Godwin's request even after Brown's retirement in 1977. Brown remembers, "[Godwin's] concern to me was [that] he wanted to make sure that we took care of the employees. I was able to keep him up to date with how MCV [Medical College of Virginia] was handling things."[57] In early December, before the full results came back, Governor Godwin established a Kepone task force to coordinate and address the sprawling impact stemming from the pesticide. James Kenley chaired the group that included Robert Jackson, Jerry McCarthy from the Council on the Environment, and members from the state's Water Control Board, Air Control Board, and the attorney general's office.

From the summer through December 1975, local and national newspapers investigated the story. One of the more influential pieces was Dan Rather's report on *60 Minutes* that aired Sunday, December 14, 1975.[58] Rather interviewed several workers affected, including Del White and his wife Pat, and Dale and Jan Gilbert. Jan Gilbert's interview was emotional. "My husband may die or he may go insane," she stated. She also opened up about sterility and the fear it generated. "[Our] children have Kepone in their bodies and the possibility that our children may not be able to reproduce or reproduce healthy children; it's more than criminal, it's just unthinkable." The video showed Nicky Shown, twenty-four, undergoing a brain scan at MCV. Dan Rather reported dramatically: "Doctors are convinced that some of the pesticide he inhaled is embedded in his brain." Shown had the highest level of any worker at Life Science—about thirty-seven thousand times the limit set by the EPA for the sewage discharge. Other parts of the report showed workers shaking and trying to hold items like a screwdriver. Rather asserted, inaccurately, as it would become clear, that Hopewell "never reported" the sewage problems to the state or the EPA. Rather interviewed Hundtofte and Moore too. They emphasized the "meager amount of information" they had possessed on Kepone's toxicity when Del White came to them about symptoms he possessed. White remembered them both saying that "there was nothing [in the plant] that would hurt a human being." "They lied to me," White said. Moore and Hundtofte both denied saying that Kepone was harmless. Moore also denied ever seeing studies done by Allied in the early 1960s showing the effects of Kepone on rats and mice that included tremors. Rather also quoted an Allied spokesperson who said that Moore was the nation's leading expert on Kepone. Moore denied that he was the expert. Rather asked if Moore was "totally knowledgeable about the hazards of the product and the safeguards necessary to produce it." "I would say no," Moore replied.[59]

Days after the story aired, and with increasing attention and scrutiny over the contamination and poisoning, the final results came in on Kepone contamination in the environment. After reviewing the results, the Kepone task force recommended that Governor Godwin close the James River, and on December 17, 1975, he did so. As he told the U.S. Senate subcommittee in 1976, "To close this great and historic river was indeed a drastic step, but I felt the public interest required action forthwith. And I could do nothing else."[60] In issuing the order, Godwin and his staff created an early narrative about the pesticide, its manufacture, regulatory issues, Kepone's effects, and the governor's response to the problems. It read in full:

> I want to reassure the people of Virginia as to the facts about the chemical Kepone, formerly manufactured in Hopewell, in order that they may be relieved of any unnecessary concern about the possible hazards involved.
>
> This chemical was manufactured for more than a year under contract to a highly reputable national chemical company, before it was brought to the attention of State health authorities as a possible health hazard.
>
> I am advised that at that time, Kepone was not subject to any State or Federal law so far as its manufacture and distribution were concerned.
>
> However, last July, a Hopewell physician reported that one of his patients who had formerly worked for Life Sciences Corporation, the manufacturers of Kepone, was apparently suffering from Kepone poisoning.
>
> Health Department checks discovered other plant personnel apparently were affected, and took steps immediately designed to close the plant and remove the hazard.
>
> All 149 plant employees have been visited and found to be exposed, in addition to some 15 others in their immediate families. Of these, 29 have received hospital treatment.
>
> It is my understanding that all of those who have shown any symptoms at all have been exposed to *massive* doses of Kepone."
>
> There is no evidence that anyone other than those who worked in the plant or their families have been exposed to this chemical in any harmful amount.
>
> Further, there is no evidence that the presence of this chemical in small amounts in the general environment constitutes a significant health hazard to the general public.
>
> All applicable State and Federal agencies have been taking emergency action to combat the problem.
>
> Water supplies for the City of Hopewell have been constantly checked and no trace of the chemical has been found.
>
> Concentrations of the chemical in sewage disposal areas and other collection points have been identified and steps taken to isolate them and eventually render them harmless.

Medical authorities advise me that minute amounts of the chemical in air samples in the past constitute no known hazard.

However, samples indicate that the chemical has found its way into the James River and some of its tributaries and into some of its aquatic population.

Therefore, as a precautionary measure, I have authorized the State Health Commissioner to close the James River from the fall line in Richmond to [the] Chesapeake Bay for the taking of *fin fish*."

The James has already been closed for the taking of shellfish by the Health Department by reason of excessive sewage discharge, and I am directing that this order will remain in effect until we are satisfied there is no hazard to public health.

Also closed to the taking of both shellfish and fin fish are the tributaries to the James River.

Let me stress that there are precautionary measures only.

Not much is known about the chemical Kepone. What is known does not tell us whether the small amounts found in fish and shellfish samples constitute a health hazard and to what extent.

Until we can be reasonably certain that no health hazard exists, the rivers and streams I have outlined will remain closed.

At the end of the statement Godwin added that he had asked Otis Brown to check state laws and "recommend action . . . for submission to the approaching session of the General Assembly which will prevent similar situations from occurring in the future."[61]

The statement's opening showed Godwin's concern about protecting Allied Chemical—not mentioning the company's name but calling it "reputable." He did name Life Science directly, though, trying to put the major blame for the crisis at that moment on LSP—he and others had yet to learn Allied's role in polluting the James. The remainder of the message was contradictory: closing the river signaled a major public health hazard, yet Godwin also tried to reassure the public. Godwin first made the case that only large amounts of Kepone exposure could cause harm by emphasizing that LSP workers were exposed to "massive amounts" of Kepone, and he said that no evidence existed that others in the area were exposed to "harmful" amounts. Yet the amounts in fish and shellfish were very small compared to amounts in the LSP workers—action levels of .3 or .4 ppm for marine life, and those relatively small amounts prompted the order closing the James River. The release only added to the confusion by asserting that there was "no evidence that the presence of this chemical in small amounts in the general environment constitute[d] a significant health hazard." He added that not much was known about the chemical and that what was known did not indicate whether "the small amounts found in fish and shellfish samples" constituted a health hazard. Yet Godwin followed the advice of his task force and closed the river. Those skeptical of Kepone's effects and the necessity of closing the river would soon exploit such uncertainty. The emergency order issued on the afternoon of December 18 outlined the rationale for closing the James and provided a more detailed, stronger case for the health risks of Kepone than did the governor's press release. Closing the river was an example of the precautionary principle as would later be outlined by the Wingspread Conference in 1998: "When an activity raises threats of harm to human health or the environment, precautionary measures should be taken even if some cause and effect relationships are not fully established scientifically."[62] This stands in contrast to what *Precautionary Politics* authors Kerry H. Whiteside and Robert Gottlieb describe as "a regulatory posture that justifies action only in cases where the scientific proof of harm is well developed and the benefits of regulation demonstrably outweigh the costs." A precautionary approach applies, they say, "especially in situations of environmental risk where by the time unambiguous scientific evidence of a serious problem becomes available, the danger may already have materialized and perhaps become irreversible."[63] In the words of environmental historian Nancy Langston, in "complex systems where consequences are unpredictable, the burden of proof should fall on industry" to demonstrate that toxic substances are safe before they are released into the environment.[64] Whether intended or not, the action to close the James to fishing and other harvesting reveals an awareness of the seriousness of Kepone contamination and recognition of the limited ability of regulators and health officials to understand and manage the risk.

The Health Department order stated that the department had become aware of conditions in the James River constituting a "potential danger to the health and welfare of the citizens of the Commonwealth due to the unauthorized and unwarranted release or discharge of Kepone (chlordecone) to the environment." The order then summarized the test results showing 1 to 4 ppb of Kepone in the water of Bailey Creek, over .1 ppb in water from the Appomattox River, and .11 to .28 ppb in the James. The results showed .02 to 14.4 ppm in fish and shellfish from the James and .21 to .81 ppm in oysters and clams. Sediments from all three bodies of water showed Kepone amounts ranging from 1 to 4 ppm. The order went on to say, "Ingestion of small amounts of Kepone may lead to accumulation and concentration in fat and body organs to levels which may be toxic to humans." And further: "Preliminary reports of experiments carried out by the National Cancer Institute implicate Kepone as a carcinogen in test animals." Reflecting the general ignorance surrounding Kepone, the order admitted, "The public health effects due to the existence of low levels of Kepone in the environment are not known at this time[,] and more investigation, research and evaluation are needed to determine whether or not health effects may endanger the public health and safety." Also: "The potential health effects of the Kepone in the flesh of fish and oysters are not fully known[,] and more investigation, research and evaluation are necessary because they may represent potential danger to the public health and safety."[65]

The ban affected an estimated five hundred commercial fishers and oyster tongers, costing an estimated $100,000 each day of the ban.[66] The ban also affected marinas and boat launching sites and affected thousands of recreational fishers. Following the Code of Virginia, sections 32–6 and 32–12 that authorized the Health Department to take action in the name of public health, the order made a misdemeanor out of the "act of fishing, catching, netting, or taking of fish by any means from the James River and all of its tributaries from the fall line to its mouth."[67] According to early newspaper accounts, fishers on the James quickly criticized the ban. "Closing the river is not going to keep me from eating," said commercial fisherman Charles Tinch of Charles City County. The Virginia Marine Resources Commission had called him to notify him of the ban. "Since I was a little boy, I have always fished during the winter. It's the way we have survived." These commercial operators were assigned specific parts of the river using stacked nets that could not move.

Reporters offered a grim, fearful set of stories on Kepone and the affected workers at the end of 1975. One came from the *Harrisonburg (Va.) Daily Record* and focused on Thurman and Jan Dykes and their children. "It's almost Christmas Eve, 1975. Thurman Dykes, 27, blinks at even the soft lights of the Christmas tree. One arm is nearly useless. He is sterile. His newborn son is ill and his wife distraught. He is a Kepone victim." The story noted how Jan Dykes had managed to get a job with Allied, but the couple had still depleted their savings since Thurman was unable to work. She revealed that some of her coworkers at Allied had been "so callous as to imitate the tremors that [had] racked both her husband and her 3-year-old daughter." Jan said that she and Thurman were mainly worried about their future and that of their children. "Who knows," Thurman said, "I may be dead in a year." They feared that their children might fall ill to the effects of the chemical. "They may never be able to work. They could be sterile."[68] The fear, psychological damage, health damage, and uncertainty of Kepone and its effects were all too real for those most affected, like the Dykes family.

Another story came from Richmond radio station WRVA. "Kepone: A Portrait in Abuse," aired on December 3, was written and narrated by reporter Neil Cotiaux who, with other station reporters, had investigated Kepone. The program outlined the conditions at Life Science, Jackson's visit, the testing, and the closure of the James River. It also summarized some of early reactions, and hinted at the deeper, more troubling information to come. Key messages of ignorance as well as trust in corporations and government came through the program. Arthur Lane, who was the former city manager of Hopewell, stated that neither he nor the city had previously heard of Kepone and that Hundtofte assured him and others Kepone was harmless. "These are private corporations, and we assume they know how to handle it," he told WRVA. Lane also said it was hard to answer whether Allied or Life Science should have let the city know about the health effects, since these were "corporate secrets" and proprietary. In a later interview, Cotiaux remembered some of these general feelings and attitudes at the time as the story unfolded. It's worth quoting at length. While those such as Dennis Treacy and David Paylor recalled the impact of works like *Silent Spring* on their own environmental consciousness in the 1960s and early 1970s, Cotiaux's thoughts indicate that, at least before Kepone, the era's national debate on pesticides and hazardous chemicals and their corporate creators may have had less of an impact in Hopewell.

> There was this mentality in Hopewell and elsewhere, inside and outside of Virginia, that if you were a local official you just trusted local manufacturers to do the right thing. They were the experts, and they, until proven otherwise, were given ... a lot of free rein to find the right way to handle products, and so that relative laxness is what helped create the Kepone catastrophe. It simply was a way of life and a general attitude that if you were a corporate official you were charged with doing the right thing, and you were expected to do the right thing, and if you were a municipal official, unless you saw a red flag, [the attitude was] "We just trust you."
>
> The general view was, in Hopewell, if you were a worker, you also trusted your employer to do the right thing. The people in Hopewell, they were like many other places, good God-fearing, honest people trying to make a buck. They would get in their pickup, go to the factory, whatever factory it was. They would work all day, take lunch hour, their lunch pail, go home, put some food on the table for their kids, and that was that. Kepone was not a household word, and everything was quite routine until Life Science Products ran amok, and that's when the alarm bells went off and people in that community said "What's this all about, and why are we being scrutinized?" It was an age of innocence, to a large degree.
>
> The sense of trust clearly had been violated in that community. They couldn't believe it was happening. How could this product be manufactured for so many years by Allied, and then more recently by Life Science Products, and managed by two former Allied officials who simply took over the production, and ... go off the rails so rapidly? And then slowly, through a lot of investigative work, primarily at the state and federal levels, the onion was peeled, and you saw that the

problems predated Life Science Products, occurred on Allied's watch, and were hidden from view for a decade.[69]

Indeed, local, state, and federal officials, journalists, writers, political leaders, lawyers and judges, corporate officials, scientists, as well as watermen and industrial workers all began to peel away at the "onion" of Kepone, discovering a longer, more troubling history—the extensive chemical relations of Kepone's creation, manufacture, and residues left behind. The initial discovery of Kepone in Hopewell launched state and federal legal charges against LSP, Hundtofte, and Moore, and a federal grand jury investigation would come as well. Legislative action would begin in Hopewell, in Virginia, and reach the federal level to address both Kepone specifically and toxic substances more generally. Studies of what to do about the Kepone waste in the river and Hopewell began. Both the EPA and OSHA reviewed their own procedures as they began to learn from Kepone. Scientists and medical authorities launched new studies and tests and published articles. Workers at LSP sued. Fishers and others affected by the ban challenged Godwin and the science, and some sued LSP. Allied also had to respond to its role in Kepone contamination as their own actions came under greater scrutiny. These developments provided some answers to several questions, among them: Just how deep and wide were the chemical relations for Kepone? Was it as dangerous as many feared? What did the owners of Life Science know? What about those at Allied? Why didn't authorities stop things earlier? What did workers know? What would happen to the fishing industry in the James? And, ultimately, would anyone be held responsible for Kepone's bodily and environmental damage?

CHAPTER 4

Biocitizenship and Accountability

The decisions to shut down Life Science and close the James River didn't end the Kepone disaster. Far from it. It was clear by the end of 1975 that Life Science workers suffered from Kepone poisoning and that Kepone from the factory polluted the environment. But much remained unknown, including how it happened, the impact and extent of damage for both workers and the environment, and whether the company, its owners, any of its workers, or government agencies would be held accountable. Newspapers in Virginia and some in Washington, D.C., began digging into the issue, and the House and Senate held Kepone hearings in January 1976. Meanwhile, civil and criminal lawsuits went forward as well. These investigations and cases involved multiple stakeholders, including Life Science workers and the company's owners, watermen and others in the seafood industry, Hooker Chemical, Allied, the City of Hopewell, the State of Virginia, and the federal government.

This chapter and chapter 5 reveal subterfuge, apathy, and missed opportunities to address the poisoning of workers and Kepone pollution. These activities, the records they left behind, and the various decisions that stemmed from them are more residues of Kepone. The inquiries and legal proceedings also set precedents and laid the groundwork for residual governance related to toxic substances. Through the residues of Kepone in the bodies of workers and residents and bodies of water, Hopewell became a "site of convergence" that showcased the links between the local and the global at the heart of chemical relations.[1] These links reveal the costs involved for local people and communities at the heart of the global pesticide trade. The investigations and legal cases are also reminiscent of what Christopher Sellers said about the knowledge of industrial hazards: "Questions about knowledge ineluctably engaged questions about responsibility." Pinpointing responsibility involved questioning the knowledge and actions of not only company executives and managers but also workers, doctors, engineers, government officials, investigators, and the lay public as well.[2] At the same time, the flip side of knowledge is ignorance. "Agnotology" is a recently coined term that refers to "the study of ignorance making, the lost and forgotten." As Robert Proctor and

Londa Schiebinger state, ignorance can be "produced or maintained in diverse settings, through mechanisms such as deliberate or inadvertent neglect, secrecy and suppression, document destruction, unquestioned tradition, and myriad forms of inherent (or avoidable) culturopolitical selectivity."[3] Much about Kepone reflects "knowledge gaps" regarding its effects and use. Some individuals hid knowledge of Kepone. Certainly neglect, secrecy, suppression, and apathy characterized the response by various government officials and managers and experts at Allied and Life Science. People and bureaucracies make decisions based on what they know, rather than what they do not know, and those involved with Kepone often acted without complete knowledge of the pesticide.[4] This was true when Governor Godwin heeded his Kepone task force and closed the James to fishing and harvesting, having little knowledge of Kepone's toxic effects nor how long the substance had poisoned the river. Workers acted without much if any knowledge of its effects. Others hid their actions in violating laws and endangering people and the environment. We can now add Kepone to the list of health and environmental subjects studied under the heading of agnotology, alongside climate change denial and the tobacco industry's manufacture of doubt on the health effects of cigarette smoking.[5]

The Social Aspects of Biocitizenship

The workers who made and handled Kepone are part of what David Weir and Mark Schapiro called the "circle of poison," the first line of people exposed to the damaging effects of pesticides. Add to them those who clean up toxic spills, are exposed to hazardous wastes, who administer the pesticides, and who consume them.[6] These are the chemical relations created from the development, manufacture, distribution, use, and consumption of pesticides. Another thing to consider is that when Life Science workers sought restitution in the courts, they invoked a form of biocitizenship, rights claims based on harm to their bodies stemming from industrial exposure to Kepone. Life Science held biopower over its workers, but workers challenged that power when their health became compromised.[7]

In considering this form of biocitizenship, it is also important to note how class, gender, and race interacted to influence how government entities responded in the Kepone crisis, as compared to other toxic events. We can't say how the story would have played out in Hopewell had the Life Science workers been African American or female or both. Whiteness and masculinity created a cruel irony for these men: it opened doors to jobs at Life Science that were otherwise closed to black men and women. Yet those benefits exposed them to Kepone poisoning.[8] As Michelle Brattain has written in her study of textile workers in Rome, Georgia, during the Jim Crow era, working-class "southern whites relied on race to serve as their entrée to politics, jobs, and, later, union jobs."[9] Race was often an "implicit facet of all aspects of southern culture," deeply embedded in "social relations, politics, and class formation." Textile mills hired whites almost exclusively, through formal and informal ways. The connection between work and status did not end abruptly with passage of civil rights legislation, as white men continued to dominate the more prestigious positions within the industries.[10] A similar situation appears to have developed in Hopewell. Because segregated labor systems remained into the 1970s, it was white male workers who were poisoned. Black residents living in the industrial sections of Hopewell near the Life Science facility avoided acute exposure but were among those who breathed in Kepone dust when it blew away from the factory. Even low-level exposure to organochloride chemicals can be hazardous. Black residents had other environmental and health issues as well. They battled the city—and lost—over the location of the landfill adjacent to their neighborhood and fought to integrate Hopewell in the 1950s and 1960s. But in terms of the Kepone story, whiteness, born from slavery, meant that African Americans remained largely outside the worst effects of Kepone. It is also notable that because of gender norms, Life Science hired men for industrial jobs. Only one woman is known to have worked at Life Science—Janice Johnson (now Denton), the office manager, a white woman.

If other incidents from across the United States and around the world are any indication, then the likelihood of Kepone receiving the attention it did had Life Science workers been women or minorities is low. Time and again, women and nonwhites have their concerns over toxics ignored or dismissed. Episodes such as the toxic exposure at Love Canal in New York or those chronicled across the South by Robert Bullard, Steve Lerner, Bryant Simon, and Ellen Spears detail the struggles of everyday citizens, often women, lower-income people, and people of color, against corporations and often irresponsible experts and government agencies.[11] From the beginning, class bias played a role in Hopewell. When men came to their local male doctors to complain about symptoms, the physicians brushed off their complaints or accused the men of drinking or using drugs. Among the local doctors, only Yinan Chou took them seriously, and his actions finally opened the door to identifying and addressing Kepone poisoning. Dr. Jackson in the state health department also took their conditions seriously, and his actions prompted the wider attention to the Kepone crisis. Once the events became public, other state agencies began to act, as did the governor, attorney general, and state legislature. In Washington D.C., senators and representatives held hearings and eventually sponsored legislation on the Kepone workers' behalf.

While workers did not share class status with legislators, their shared white masculinity may have influenced the ability of the Life Science workers to gain an audience with Congress and how those officials responded. Kepone poisoning put at risk the fundamentals of traditional masculinity—the ability of these men to act as the sole breadwinner and to father children. In congressional hearings, Dr. Sidney Houff of the Medical College of Virginia drew on these dynamics to explain why more Life Science workers didn't demand more action from their doctors. "These people are not chemists. They are not college graduates. These are men who are used to taking orders, whether it is in service or whatever employment they have, and they are the type of people whom, when you tell them something, they believe it, and they did believe in their employers."[12] He emphasized reproduction too. "To be sterile for any man, I think, is a very emotionally devastating thing. Many of these patients are young. They have never had an opportunity to complete their families even if they started. They cannot now look forward to completing those families with any type of certainty." Workers in the hearings or interviewed by news media noted sterility, along with cancer, as their major fear. Houff went on to note how some of the men had reached out to him. "I know 3 o'clock and 4 o'clock in the morning is not an unusual time for me to get a phone call from these patients telling me they are in trouble. They are scared; they are tremoring; they are worried; they want to talk."[13] Chair of the House Subcommittee on Manpower, Compensation, and Health and Safety of the Committee on Education and Labor Dominic Daniels echoed these sentiments. "Do inadequate or outdated compensation laws doom these men and their families to a life of poverty by stripping them of any hope of financial security?" He made reproduction front and center as well. "The ramifications of chemical intoxication are of horrifying proportions. The possibility of developing cancer; the fact that exposed humans may not be able to reproduce; the evidence that Kepone can be passed to offspring; and the untold mental anguish of workers, their wives and their children must be thoroughly investigated."[14] Tom Fitzgerald sensed something shift at the House hearings held in Hopewell High School when it came up that Del White had been in the Marines. When a representative "found out that Del White was an ex-Marine, which he was also, there was something about that that changed his tune. He went from 'maybe everybody is doing their best to tell the truth' to 'these management people are a bunch of lying sons of bitches.' *Semper fi*, man."[15] Fitzgerald likely remembered Congressman John Dent, Democrat from Pennsylvania, who praised White as a fellow Marine. "In the Marines, ... we were taught ... never to lie and never to walk backwards. I am happy that you don't lie, and I see that you are not walking backwards. I thank you very kindly."[16]

Those wives who were interviewed at the time indicated that it was their role to manage the home, and investigators assumed this. Their questions to these women about Kepone focused on finding it in the laundry and about any medical effects on their children. The wives also confirmed the physical and mental effects of Kepone poisoning on their husbands and described the impact on them and their children. Yet they could also be strong advocates for their husbands and the family livelihood. Jan Gilbert, for example, pushed her husband Dale to see Dr. Chou. She, Pat White, and Jan Dykes opened up about the effects of Kepone to reporters and investigators. They answered questions in hearings, providing details of Kepone's effects, the amount of medical bills they received and couldn't pay, and the testing they, their children, and their husbands underwent. In these ways they affirmed their own and their husbands' legitimacy, their own fears, and their desire for restitution for what happened.

Opening Salvos: Ignorance and Accountability

Documents from legal and congressional investigations help reveal the developing knowledge of Kepone and accountability for the damage done to workers. In September 1975, as publicity about what happened at Life Science increased, a group of ten former Life Science workers—including Dale Gilbert, Del White, and Frank Arrigo—filed a $24.9 million personal injury suit against Allied and Hooker Chemical (*Gilbert v. Allied Chemical*). They could not sue Life Science directly because, under Virginia law, workers receiving workers' compensation from a company cannot sue that company.[17] It was the first of the Kepone legal proceedings. Asking for between $2 and $3 million in damages per person, the workers' counsel alleged that Allied and Hooker, which supplied the materials for making Kepone, failed to provide Life Science or its employees with adequate warnings on the toxicity of Kepone or its ingredients.[18] Other major cases came forward in 1976. Janice Gilbert was the lead plaintiff on a lawsuit for some $55 million on behalf of family members of the Life Science workers, including children. Two others were filed on behalf of former Life Science workers not part of the *Gilbert* case, including Jerry Collins and James Moore. Other Life Science workers filed their own lawsuits. Most were consolidated into the *Gilbert* case and some into what became the *Collins* case. Janice Gilbert's case remained on its own. These three main cases involving some fifty-six workers, as well as wives and children, sought $108.9 million in damages.[19]

Yet another suit came forward from two railroad workers claiming damages from hauling the ingredients to and from Allied and the Life Science factory. Then, in January 1976, U.S. Senate and House hearings on Kepone added to the growing investigations. The hearings were aimed at uncovering the role of government agencies in the unfolding events and deciding whether to amend or create new legislation related to worker safety and toxic substances. They also brought to the fore several key issues, including the degree of toxicity of Kepone and its health effects on humans, whether Allied or Hooker Chemical knew this information, and whether Hundtofte and Moore were aware but failed to adequately warn or protect their workers with equipment and safety training. These questions of corporate and individual responsibility combined with questions of local, state, and federal government accountability.

Lawyers for the workers in the *Gilbert* case first issued an interrogatory to Allied to establish the facts on which the case might be tried. The complaint alleged that both Allied and Hooker "knew or should have known that" Hexachlorocyclopentadiene (HCP), sulfur trioxide, and antimony pentachloride are "imminently and inherently dangerous to life or property," yet both "negligently supplied said chemicals to Life Science without notice of warning of the defect or danger to the plaintiffs who were users of said chemicals." They argued that both Life Science and Allied failed to take reasonable precautions against the danger and that Allied failed to require Life Science as a contractor to take necessary safety precautions.[20] They sought to get Allied to admit as statements of fact that Kepone and the ingredients used to make it are "highly toxic to man" and that Allied had this knowledge. They claimed that Allied knew of results in animal testing done in the 1960s but did not warn employees of LSP about these results that included tremors, other symptoms the workers showed, and possibly cancer. Neither the shipping labels for Kepone nor the rail cars supplying HCP, SO_3, and antimony pentachloride showed the common skull-and-crossbones symbol or had POISON in prominent letters. They also noted that Kepone was not registered as a pesticide and that LSP was not registered to make it.[21]

Allied offered a vigorous rebuttal. In general, the company denied its liability, asserting that the workers or others at Life Science were negligent in handling Kepone and other substances. Lawyers admitted that, of course, Allied knew of the tests but refused to admit that the company had a duty to warn Life Science employees. Allied had supplied shipping labels for Kepone that contained clear warnings about it being fatal if swallowed, inhaled, or absorbed through the skin. The labels also said to avoid getting it in the eyes, on clothing, or in food. Allied lawyers also pushed the responsibility onto Hundtofte and Moore, who, they asserted, were "knowledgeable about the properties of Kepone and in the industrial safety measures to be observed in producing Kepone."[22] Since Kepone was not labeled as "highly toxic" under federal regulations, there was no need for them to have put visual signs of toxicity on labels. According to EPA regulations, neither Allied nor Life Science were required to register Kepone as a pesticide—they claimed it was only an ingredient in making one. That requirement did not change until August 1975—after Life Science had closed. The company also fought back against the stated health effects. Lawyers argued that the phrase "highly toxic" was "vague and imprecise" and "argumentative," as were the phrases "excessive bleeding," "impaired liver function," "cause sterility," and "cause loss of body weight." Allied also argued that it was not known that Kepone could "cause the development of cancer."[23] There was inconsistency in Allied's argument. A *Washington Star* report on December 27, 1975, quoted Allied spokesman Norman Herrington stating that the company was "very much distressed" about the poisoning and contamination. But he added, "Legally we don't feel we are responsible." Herrington stated that Hundtofte and Moore knew Kepone "was toxic" and said "Allied has never concealed the fact that this material was highly toxic."[24] The story also featured Paul Larson, the scientist in charge of Kepone tests at MCV in the 1960s, who argued that Allied knew then that Kepone was highly toxic to animals and therefore humans.[25]

The case file included an evaluation of Kepone by Samuel S. Epstein, the noted scientist and writer on environmental health who the *New York Times* would call the "Cassandra on cancer" in its obituary for him in 2018. At the time of the Kepone crisis he was working on one of his most controversial books, *The Politics of Cancer*.[26] Epstein reviewed existing research, especially the studies done by Larson and others in 1960–62 that showed liver cancer in rats; the same results emerged for mice and rats in the National Cancer Institute studies done in 1976. He concluded that these results established "the probability that Kepone [would] also be carcinogenic to exposed humans."[27] A second report came from Rudolph J. Jaeger, then in the School of Public Health at Harvard, who would go on to research widely in toxicology and environmental health. Jaeger concluded, "Kepone is a truly chronic toxin more potent than many other chlorinated hydrocarbon insecticides." He said that Life Science workers were likely to develop cancer, given their exposure.[28] In long letter to *Science* magazine in July 1976, Jaeger also went on to criticize the testing and evaluation process at the NCI on chemical carcinogenesis. He noted that the director of this program at NCI, Umberto Saffiotti, had recently resigned because of the lack of staff assigned to promptly assess and disseminate experimental results. Kepone illustrated those problems. Studies of Kepone began in 1971 on mice and 1972 on rats, but results were not released until April 1976, well after workers had been poisoned. The NCI had already completed its bioassays by the time Life Science opened for business in March 1974. If both studies had been evaluated in a timely way, the results would have been available in January 1975, possibly giving authorities and medical professionals knowledge to address workers' safety sooner. Jaeger also noted that the earlier tests done in the 1960s showed the chronic dangers of Kepone.[29]

The more open-ended discussions at the Senate and House hearings in January 1976 probed how much was known about Kepone (and by whom and when), conditions inside Life Science, the relationship between Allied and Life Science, and the degree of knowledge and involvement of local, state, and federal authorities. Much attention was paid to Hundtofte and Moore, who both appeared at each hearing without legal counsel and answered questions from senators and representatives. As expected, Allied executives sought to distance the company from Life Science. According to Allied's G. C. Matthiesen, president of the Agricultural Division, both Moore and Hundtofte were "experienced in the operation of... Kepone production at Hopewell." As assistant director of research for the Agricultural Division, Moore "had been in charge of Kepone during the first three years of production" and was "familiar with the procedural and safety procedures for the process." Moore was credited with "design modifications which led to a significant increase in the Semiworks capacity in 1971." As for Hundtofte, he had been the plant manager for the Agricultural Division plant in Hopewell, "a facility with almost 400 employees," and responsible for several years for the production of Kepone in the Semiworks. This was after Kepone processing was transferred "from Moore's research group to the manufacturing operations which were part of Hundtofte's responsibilities." According to Matthiesen, Moore and Hundtofte "knew as much or more about Kepone production than anyone in Allied Chemical, or probably in the country."[30] How much of this is true and how much reflected Allied shifting blame became a matter of debate.

Both Moore and Hundtofte downplayed their knowledge at the Senate hearings. Moore said: "I would hasten to say that my interest has been research and development and not in Kepone or its production." Senator Patrick Leahy (D-Vt.) asked him about his Allied experience: "Weren't you in charge of the Kepone?" "In no way," replied Moore. "Some of the people reported through the research organization eventually to me just as they eventually reported through the vice president, but I definitely was not solely in charge of the Semiworks." In response to Matthiesen's claims, Hundtofte stated: "That is not a fair statement with regard to my qualifications on Kepone." Hundtofte said he never even knew all of the steps necessary to make Kepone until he met with Moore to start Life Science. His main expertise was in ammonia production, which occupied most of the Semiworks; Kepone was only a small amount of that operation.[31]

A key issue of dispute was the degree to which Moore and Hundtofte were aware of Kepone's toxicity—and how they handled it at Life Science. Allied tested Kepone beginning in the early 1960s and maintained a file on the results, a set of documents that came to be known as the "blue books." Despite their claims, both Moore and Hundtofte were at some level in charge of Kepone production for Allied. During both the House and Senate hearings in January 1976, members of Congress asked them if they had access to these results and were aware of Kepone's toxicity. Both demurred on this point. Senator Leahy asked Moore: "When did you first become aware of the total toxic nature of Kepone?" He replied: "Well, I do not know if I am totally aware now. It seems to be more toxic than I am aware of. Democratic senator James B. Allen, chair of the committee, asked Moore about the blue books. "Did they furnish you with a blue book showing the toxic nature of Kepone?" Moore said no.[32] This line of questioning returned when the House of Representatives held hearings at Hopewell High School in January 1976. These were part of the Committee on Education and Labor's role in oversight of the Occupational Safety and Health Administration (OSHA). Dominick Daniels (D-NJ) chaired the hearing. Both Moore and Hundtofte claimed that they had only seen these toxicology studies in August 1975, after Life Science closed down.[33] Yet these studies were subpoenaed from Life Science as part of the investigations. In the Senate hearings, Moore backtracked a bit when presented with that fact. "If that was in there, I am certain I saw it." He saw it when he was looking through everything for the subpoena.[34]

As contradictory as he was, Moore was possibly truthful in asserting Allied had not provided him with the full testing results. Moore would later file his own lawsuit against Allied in 1977. In it he maintained that Allied had failed to supply him with complete toxicity information and concealed Kepone's true dangers on humans and the environment. He also maintained that degrees in chemistry and chemical engineering do not "automatically confer knowledge of toxicology, pharmacology or industrial hygiene." In a sworn deposition for the case, even one of the inventors of Kepone, Dr. E. E. Gilbert, admitted that he only had a general knowledge of what "LD50" meant, and he said that his knowledge of toxicity testing was "rather sketchy."[35] In toxicity, LD refers to the "lethal dose" of a substance. So LD50 means the lethal dose of a substance, in this case Kepone, given at once that will kill 50 percent of the test animals. Toxicologists test to determine toxicity, Gilbert insisted, but chemists who make the product do not. Moore made the same argument, noting that Allied used another toxicology expert, Dr. W. A. Knapp, for its testing. If one of the inventors had little knowledge of the toxicity of the invention, how could Moore have known? While at Allied, Moore worked in areas other than pesticides and in research on new chemicals and processes, not Kepone. He was "certain he was never in Allied's Hopewell Semi-Works when Kepone was being produced." He even tried to tour the Semiworks "just before Life Science contracted with Allied and was not allowed to do so." He further swore that he asked Allied for toxicological data and to see Dr. Knapp, "all to no avail." What information he did obtain was about Kepone's efficacy, about its use in small amounts in bait, and other documents from Allied that stated clearly that Kepone "was not particularly hazardous" or that "Kepone is among the less hazardous insecticides to make and use." Even Allied's director of occupational health and product safety, Warren Ferguson, was unaware of the blue book studies from the 1960s. Knapp only shared those with him in September 1975 as investigations were underway. Further, Knapp stated that he prepared the blue books on toxicology studies on animals from the 1960s from his file on Kepone at the Allied office in New Jersey. He first sent the books and other data to Ike Swisher of Allied in Houston in July 1975 with a letter stating that he had no objection to sending these to Life Science, provided that Allied "accept no liability for their completeness or accuracy" and on a "confidential basis with no public release except to regulatory agencies."[36] The data submitted had never been published. Allied's attorney in Houston forwarded several of these documents to Life Science in late July or early August, after the factory had closed and for use by Life Science in dealing with the Virginia Health Department. Critically, Allied lawyers kept the blue books and toxicological studies of Kepone on quail, ducks, and crabs. What Life Science received were safety and handling procedures, information on hazards of inhaling Kepone dust, a list of studies, and product technical data.[37] For his part, Hundtofte insisted that the set of data in the blue books was in the possession of the Health Department, not Life Science.[38]

Both Hundtofte and Moore continued to use ignorance of Kepone in their defense, deflecting responsibility elsewhere. They qualified their knowledge of Kepone's toxicity by insisting that they relied on the LD50 rating in understanding Kepone. The LD rating is in milligrams per kilogram of body weight and useful for measuring acute toxicity, but that is not necessarily what happened to the workers at Life Science. They occupied a space that meant less than an acute exposure to a large amount of the pesticide but not a chronic exposure over years. Those who worked there were exposed for weeks or months, pushing them away from immediate exposure in one brief dose (like swallowing or inhaling vapors in a small amount), to something like a chronic situation. The lower the LD50 number, the more toxic. Kepone's number is 126; DDT is 250. Kepone was often compared to DDT, which also shaped how Moore and Hundtofte understood the substance their company made. As Moore stated to the House committee: "We had about 10 pieces of literature relating to the toxicological work on ... a variety of test animals, which, to my understanding, related to LD50's or lethal dose. To my knowledge, my understanding is a better word, it placed Kepone about where DDT is in toxicity in two orders of magnitude less than toxic, amongst things like parathion which is a toxic pesticide."[39] Once the conditions of the workers came out, Moore reflected, "I think we all recall the pictures of troops putting DDT all over a populace when they moved in." "I think we now know the evidence[,] which [shows that] this is a bad way to evaluate on the basis of LD50." Hundtofte had a similar view. "Unfortunately, the sort of understanding that I had about toxicology was in terms of how much it takes to kill something. This is really referred to in terms of the LD50 data that we had. This is a very limited way, unfortunately, as we know in hindsight and an entirely wrong way to look at toxicology. I had some of my own experience in a woeful upbringing of using DDT and I suppose that had some influence on me."⁴⁰ Moore admitted later in his own lawsuit against Allied that "he would have never participated in Kepone production" if he had known then everything he had since come to know.⁴¹

Senator Allen pushed Hundtofte hard on the issue of toxicity. "Well, I assume since you and the plant manager were at Allied in the production of Kepone, you knew about its toxic qualities?" Hundtofte replied: "No, frankly, I did not." Allen remained perplexed. "Well, would not your curiosity have been peaked [sic] a little bit if you are going to [be] working on one product, would not your curiosity be aroused as to what in the world this product is and what effect it might have on people working with it?" "Well," Hundtofte replied, "I guess that this is one of those situations where I relied on what experience that I could see, and this is, at that time there was no knowledge of any toxic effects. I worked in the plant myself. We started producing in March 1974, and I had no other supervision between me and the operators until essentially July of 1974. So, I had that experience. I was not aware of any effects at that time." He later stated, "The real tragic thing in this is we did not know the degree of toxicity that we were really coping with." So, Senator Allen concluded, "[You] just turned these employees loose working on this highly toxic product without knowing yourself whether it was toxic or not?" Hundtofte maintained his line of defense. "Well, I certainly did not know the degree of toxicity. Again, having been into the plant myself and having operated it, I suppose I relied on that as a basis of experience."⁴² He also elaborated on how he understood toxicity as something associated with chemicals and their acute poisoning. "But again, in my experience, it has been usually a chemical that you are dealing with that would overwhelm you if you were either enclosed in vapors or in the liquid or a massive contact on your skin, and this was the sort of association that I had with toxicity." Senator Allen pressed Hundtofte on the Kepone dust: "You did not know that it would be toxic in its dusty form by breathing it into the lungs?" "No; I certainly did not," replied Hundtofte.⁴³

> I spent many long hours there. That was literally true. I think you could get this determination from anybody who was there. In order that my family could be with me, sometimes it was on a regular basis, one or two times a week my wife, my just month-old son, and a 6-year-old son would come down and stay in the office, in this area where we described the eating, and they slept. They stayed on the same cot that Del White slept on. So, when we talk about my involvement in it, it was not in a part involvement—the real tragic thing in this is we did not know the degree of toxicity that we were really coping with. So, I have been in that plant and I have some of the symptoms that are characteristic of the Kepone poisoning. I hope there is not any problem with that. I feel that it is minor. But I guess what concerns me most at this point is the fact that my wife is pregnant

and expecting, so I am not removed from the situation. Certainly, even apart from that, I could not be removed from the situation in view of what has happened to these people.[44]

Senator Leahy seemed satisfied of Hundtofte's sincerity. "I am impressed, Mr. Hundtofte, by the fact that you were actively involved in the plant and you were obviously spending time there, according to your testimony. I am willing to accept the fact that you probably did not know about the toxicity of Kepone.[45] Reflecting back on her experience working at Life Science, former office manager Janice Johnson gave Hundtofte the benefit of the doubt as well. "Why I wasn't that concerned was that Virgil Hundtofte was very much involved with the manufacturing of Kepone. He went out in the plant routinely to check the equipment, the product, everything. He would come in covered with Kepone dust. It didn't seem to bother him. He even was married, at the time, had just gotten married, and had a baby. That baby now is probably a grandfather, but he would bring the baby to the plant and put it right on the floor where people walked in with Kepone on their shoes, their boots, and he didn't seem to be concerned about it, so in my heart I don't think he ever knew the potential danger."[46]

With similar explanations for how they understood Kepone's toxicity, how aware and concerned were Hundtofte and Moore about the conditions inside the plant and the growing numbers of poisoned workers? Moore expressed his concern for the workers, but he shifted responsibility for the conditions in the plant to regulators, especially those in Virginia. "I am concerned about the people. I am vitally concerned. I would say that I personally worked more than full time for their benefit, to handle the analytical end and to develop new things for these people to do. I didn't take a vacation." In terms of what could have been done to prevent the tragedy, Moore said: "My feeling is that the people from Virginia worked as best they could with us. Whether it was lack of talent or ability, I don't know. We kept an open door to them to do anything they suggested." Senator Allen asked him: "Wasn't there any such responsibility on management to advise these people of the inherent dangers connected with the production of this product?" Moore had written the safety protocols for Life Science and stated that from his perspective that was enough. "From my standpoint, I think that was certainly done in the safety instructions and the process description." "I understand," he went on, "that those were followed as the standard operating procedure." In writing the instructions for Life Science, he used the set from Allied. Moore claimed in the Senate hearings, "The instructions I wrote, I believe I mentioned in the first paragraph said: Kepone is a toxic insecticide."[47]

That was not the case. In fact, there are critical differences in what Moore included and what Allied provided. From the Life Science instructions: "Kepone and the materials used to produce it should be handled carefully to prevent swallowing, absorption through the skin, or getting in the eyes. Each of these materials can be hazardous if improperly handled although none is a highly poisonous substance. Further information on handling safety is attached to this memorandum." Allied's original instructions, written by Jameil Ameen, were much more thorough.

> The raw materials and desired product of this process are very toxic or reactive materials which must be handled carefully and kept from the skin, eyes, respiratory system, and mouth. Neoprene gloves, rubber shoes, and face shield must be worn when handling these chemicals and adequate ventilation maintained to remove vapors and dust. Material contacting the skin must be immediately washed off with soap and water. Material entering the eyes must be flooded immediately with water, washed for a minimum of 15 minutes, and medical attention immediately obtained. Material taken orally should be immediately removed from the stomach by inducing vomiting and medical attention immediately obtained. Operators should shower thoroughly after work before dressing in street clothes.[48]

One item missing from Allied's instructions was the requirement for respirators; it seems even Allied workers did not have those until 1970, four years into their Kepone production. Some Allied workers who made Kepone may have had some ill effects as well, although evidence for this is lacking.[49]

Moore stated that the first he had heard of an issue in the plant was in the summer of 1975, right before the plant was closed. Moore didn't spend much time in the plant; he limited his visits to the office to two or three times per week. For most of Life Science's existence, he spent his time in a lab some twenty-five miles away. Moore was physically closer in March 1975 as the company put a down payment on property about a mile from the Life Science plant and moved the office to a trailer on that property. He and Hundtofte were planning on expanding into a new chemical venture. Moore never inquired about or seemed to know about the high turnover in the plant.

Moore and Hundtofte both denied the allegations from Del White about an encounter the three men had in the Life Science office over White's health problems related to Kepone poisoning. Moore claimed that White said he felt nervous and wondered if Kepone could be the cause. They went through the materials on Kepone in the office with Hundtofte. "I felt that he was convinced—in fact, my recollection is that he said, that is not what I have got. I also specifically remember that Mr. Hundtofte said something to the effect that you should wear—you should stay clean enough that you could wear a blue serge suit, in the performance of your duty. He was advised to go to a doctor by Mr. Hundtofte at that time."[50]

Moore denied ever saying Kepone was "not harmful to human beings": "This can only be rather strongly denied; and I say this is absolutely not true." He remembered referring again to the LD50 number and looking over the materials in the office: And the indication taken by Mr. White was: Well, this is a stomach poison and this is not what is bothering me."[51]

Hundtofte also denied the allegations. "I would further like to say that in that meeting there was no comment made, and I stress no comment made, that anything that is out there would not hurt you. I do not think anyone can work in a chemical plant without having some degree of restraint for whatever material it is that they are handling. Mr. Moore and I were both there. I advised Del White that if he had any concern, any reservation, about his health, to go to the doctor. We advised him that we were not aware of any effects on human beings regarding this material. Of course, I had related my own experience of working in the plant."[52] He even asserted that in April 1975 White had refused to go to the doctor, even after Hundtofte made the appointment himself.[53]

Unlike Moore, Hundtofte worked often in the plant, at least until March 1975 when Life Science hired H. D. Howard to manage operations while Hundtofte focused more on developing new products with Moore and spent time at the trailer office a mile away. In terms of safety, Hundtofte assured the Senate and House subcommittees that safety was important for him and the workers. Since they were a small firm, unlike Allied there was no "safety department," but he added, "I would say with my experience and background, there is little doubt that we discussed safety on numerous occasions." He observed his workers using the safety gear issued to them. "There was a standard procedure for a new employee where he was issued a hard hat, glasses, soft-side gloves, neoprene gloves, and later a respirator. In the early stages, we did not use a respirator."[54] There was training for employees as well. "We provided training for our employees. We usually devoted 2 to 3 days on one part of the plant, and on the other part of the plant devoted about a week or more to that kind of training." Hundtofte admitted that sometimes even he didn't follow the safety protocols. Workers would take off their gloves and pack the bags full of Kepone and then tie the bags shut. "Unfortunately, I am one of those employees."[55] He often would sample the Kepone with his bare hands as part of his inspection before shipment. Even Allied observers would come in at times and do the same.

Although he was in the plant, Hundtofte was unaware of workers with severe tremors or other medical problems stemming from Kepone poisoning. He didn't know there were any issues like these until he spoke with Del White in November 1974 and then again in April 1975. He recalled speaking to Dale Gilbert, but that was about weight loss and not other symptoms. The only medical issue of which he was aware were workers with rashes, for which the recommendation was using Vaseline. Besides White and Gilbert, no one told him about their other conditions. "I spent long hours there at that plant, Del White and I worked very closely together, as I did with Gilbert and Bill Moyer, and the other shift supervisors, and we spent long hours together, and not once did any of those people come to me, other than a description of feeling nervous, and this was with Del White." Senator Allen asked if White reported that others in the plant were having medical issues. "No, he did not," was Hundtofte's reply. "I can refer specifically to Thurman Dykes, who was here and testified, and I talked to him on numerous occasions, and I did not recognize any shaking nor did he ever say anything to me about a shaking problem." The only complaints Hundtofte recalled were "about working conditions... in regard to having to walk up three flights of stairs to check an operating unit, to the fact of moving 200-pound drums, to the fact of having to sweep up and clean up in the operation. And that they would much prefer to work on a control board kind of complex, rather than something that had so many manual valves in it." Senator Leahy mentioned "complaints about... inadequate shower arrangements" and a welding cable on a wet floor. Hundtofte said he had heard no complaint about the shower or the welding cable in the water. He acknowledged getting protective garments only at the very end of the plant's existence (workers noted that these were disposable paper ones that Life Science sold to them for a dollar each). He insisted that the plant was clean, at least until April 1975 when they installed a new dryer that allowed for increased production.[56]

When asked about responsibility, Hundtofte was vague on his own role, put some blame on the workers, and also claimed ignorance of Kepone's effects. Senator Leahy asked in exasperation: "I mean isn't anybody responsible?" "Well," replied Hundtofte, "I would certainly feel like many people are responsible, because there have been many facets of this that different things could have been done, different decisions could have been made. I think it almost goes to the point of including our own employees."[57] He felt that Allied withheld the toxicology studies and that had he known more about Kepone he would have been able to consider some of the side effects. These studies were on file at Life Science, but Hundtofte claimed he did not see them until August 1975, after Life Science shut down. Senator Allen asked with frustration, perhaps tinged with sarcasm: "It looks as if after the fact you are learning quite a bit about Kepone, are you not, Mr. Hundtofte?" "Unfortunately," Hundtofte responded. Allen continued: "Do you not think it would have been well to have learned a little of this when you took over the job of manufacturing Kepone?" "It certainly would have," Hundtofte answered.[58] Allen commented later, "It looks to me, Mr. Hundtofte, that... there was a great amount of ignorance among those who were in charge of the production of this product, and a failure to use the information that was available. Would you say that is a correct assessment?" Hundtofte replied: "I think in hindsight it is a cor-

Frank Arrigo (*left*) and Tom Fitzgerald, both of whom worked in the production area of Life Science Products and joined the legal cases against LSP and Allied. Tom Fitzgerald passed away from cancer in 2022. This image is a still from the 2019 documentary by Bernard Crutzen, *Pour quelques bananas de plus: Le scandale du chlordécone* (For a Few More Bananas: The Chlordecone Scandal).

rect assessment."[59] Workers, of course, would agree. They were less than forgiving when it came to Hundtofte and Moore's defense of their own knowledge of Kepone, what they said to employees, and about the general safety protocols and conditions in the plant.

One reason for the lax safety regulations inside Life Science might have been the hectic production schedule. The original plant Moore designed had a capacity for the production of three thousand pounds of Kepone per day, but the plant at times reached six thousand pounds per day. "We were told by Allied that they could take all the product that we could produce," said Hundtofte.[60] This was twice what Allied had made when it produced it. As Tom Fitzgerald remembered, the plant "ran seven days a week, 24 hours." He remembered Life Science adding the second dryer to keep up the pace. "About 2 or 3 weeks before I left, they just put in a new dryer facility which increased production by at least 100 percent. So there was no worry about people being laid off and lack of work. The people that did quit, there was a rapid turnover. Very few of them left for any other reason than [that] they [could not] stand the shakes or the rash."[61] Frank Arrigo remembered that too. "And that was like 24 hour[s], seven days a week. . . . They wanted to get out as much of that stuff as they could."[62]

Steve Keavy worked at Life Science. Decades later he remains skeptical regarding Moore and Hundtofte not being aware of what was happening with the workers. At least he think they should have been. "I can't tell you what anybody knew,

BIOCITIZENSHIP AND ACCOUNTABILITY [83]

but... I knew something was wrong. I didn't know anything from anything, [but] I knew something was wrong, so there's no way you can convince me that the supervisors didn't see people. There's no way you could walk through that facility and interact with a guy like Nicky Shown and not see something was wrong, no way. I don't care who you are, how dumb you want to plead. No way they could not have known, no way." He also feels that Moore and Hundtofte had to have known about the toxicity of Kepone from Allied. How else could they have let this happen? "Allied was not a little company, at the time, and Allied was a pretty savvy operation. They had to know. They had to have translated the dangers of that product to Bill Moore and his people, and it was just ignored." Keavy also remembers the lax safety. "Safety equipment, PPEs [personal protective equipment], was absolutely ignored,... and that's what I remember the most... that nobody ever told me to wear the stuff."[63]

Thurman Dykes worked at Life Science from November 1974 to February 1975. He also worked for Allied as well. His pay at Life Science went from $3.75 per hour to $5.25 (about $27.50 in 2020). He averaged about $150 per week in take home pay. Workers were reluctant to use the shower facilities. "They have shower facilities, but not many people used it, because it was out in the open, and if you had wind... out on Randolph Road you [did] not want to go out there to take your shower." He remembers workers wearing goggles, but the dust blew up inside them anyway. They had rubber gloves, but he said, "You had to cut the top out of the rubber gloves because the Kepone would get in there and tear your wrist up." He stated that managers never emphasized the dangers of Kepone, only the chemicals used to make it. Conditions led to high turnover. "A lot of the guys came, they got laid off at Allied with me, and they were back there in the dust, and they said they could not take it. They quit."[64]

William Tatum came to work at Life Science only two weeks before they shut down. He remembered plant manager H. D. Howard talking with him. "I reported to Mr. Howard for work, and he told me that some of the workers got tremors and a rash, but he said it was temporary and it would go away. He said a whole lot of it might kill a rat or a small rabbit or something, but it wasn't permanently harmful." He had goggles, rubber gloves, a dust filter, and boots. "Anyway, I started working there, and after about 2 days I had a systemic reaction in my throat, and my eyes started burning towards the end of the day, and I was driving home, and it got worse and worse, so I went to the emergency room at Petersburg General, and they washed out my eyes, and there wasn't much they could do. It just kept on burning for about 8 more hours, and it just burnt itself out." After returning to work, Tatum said, Howard gave him "a full-face gas mask with a canister"; there was one of these that Tatum would share with others working in the dryer area. He would often sweep up Kepone dust from the floor and pack it in the barrels. There was so much on the floor he could get "about 150 pounds" of Kepone. Much dust remained in areas that weren't open. If a batch of Kepone was too wet even after going through the dryer, "the barrel was dumped on the floor." "Then," he said, "you took a shovel and put it back into the machine and [swept] up the rest of it, and if you felt like it you'd sweep up the whole floor."[65]

Del White cut an authoritative figure at the House hearings in Hopewell when he told the subcommittee he had been in the Marines from 1962 to 1966. He had then worked at Firestone and Allied before joining Life Science, becoming the plant superintendent in August 1974. "After going to work full time, I worked somewhere from 12 to 16 hours a day, 7 days a week, in the Kepone production area. I noticed after I had been there somewhere between 2 and 3 weeks, I started to develop some more of a tremor." White raised concerns over safety and health; he recalled his meeting with Hundtofte and Moore when, he insisted, they had assured him that Kepone could not be the cause. He also recalled the many others who were in the plant, and none raised the alarm. "I had seen the State officials in there. I personally took the people from Travelers Insurance on tour that were checking the plant. They mentioned that there was no problem there." Even Allied personnel appeared satisfied. "I remember one time in February or March of 1975, somewhere around that area, there were three officials of Allied that toured the plant. I was on top of a dryer when they came up. We were in a normal operating day. Everybody did seem to be happy with the way things were going."[66]

Tom Fitzgerald remembers an incident involving White that required a hospital visit.

> There was a particular afternoon...I think it might have been the day I was messing with concentrated doses of HCP, and I just had this sudden outlandish reaction to it, hyperventilating, shaking like hell. My vision was a little hinky, and I got this gigantic headache, and I told Del about it, and he went up to the office to get someone to take me to the emergency room. No one was available or willing, so he went hillbilly Marine on them and said, "Okay, I'm taking his ass to the emergency room. If you want to go run the goddamn place in the meantime, you go right ahead, but we're leaving." He did, took me to the emergency room and got my diagnoses of too much caffeine and [got] a shot of valium in the ass, and I still remember him raising hell. His face was beet red, walked right in back to the plant raging about what was going on. He knew that there was something deeply wrong there. He knew that he was being lied to, and he felt personally responsible for the damage to the rest of us. He was a good man.[67]

In February 1975 White was able to convince Hundtofte to get masks for workers but not respirators. "I initiated the dust masks that they allowed me to buy because I explained to Mr. Hundtofte that we had to buy the dust masks because the dust back there was unbearable. At that time, dust masks were purchased." He "recommend a respirator that was more expensive, but they resolved to get dust masks." White was asked if any changes were made while he was there through April 15, 1975, to improve the dust problem. "No, sir; there was none. We were told all changes were made to increase production. Everything that was done was done to increase production."[68] Del's wife Pat related her view of how things were affecting her and her husband. "He developed a rash when he was working there part time. Then it was not long, just weeks after he went there full time, that he started shaking. Then months after that, I started noticing his eyes would jump. We talked about it and discussed it. He was working tremendous hours; sometimes we would not even see him. There were problems all of the time because the phone rang. Sometimes before he would get home in the evenings, the phone would ring and they would have another problem." She and Del assumed it was just long hours. Eventually "I told him that the job and the money, it is not worth it. We discussed it and he quit."[69]

At some point, Hundtofte advised workers to use Vaseline for the rash. But that only made things worse. Fitzgerald recalls: "The instruction was to keep your face covered with Vaseline to keep the stuff off your skin, and it turned out that it helped it penetrate your skin."[70] In his testimony to the House subcommittee, MCV neurologist John Taylor concurred: "Vaseline may indeed enhance absorption since it is oil soluble, and particularly if all of the Kepone from the previous day was not removed before the Vaseline was applied."[71]

Fear and the Uncertainty of Damage

Besides detailing the issues and disagreements with Hundtofte and Moore on safety, working conditions, and knowledge about Kepone, workers and medical professionals at the time expressed their uncertainty and fear regarding Kepone's health effects. Workers with the worst symptoms and some of their family members went to the Medical College of Virginia in Richmond. A team of doctors led by Dr. Taylor treated them and also studied ways to depurate Kepone from their bodies. At the Senate hearings, Dr. Taylor detailed the effects of Kepone poisoning. Many described personality changes, mainly "irritability." "Many of them also had substantive weight loss, in one or two cases exceeding 50 pounds. A certain number also had headache, gait difficulty, and rash." Of those most affected, the "signs on examination of these people were rather striking. Nineteen of the 21 had a clear-cut tremor. Many [had] loss of memory." Taylor and his team called the eye movements "opsoclonus" and said it "clearly implicated the brain as a problem in these particular patients." Taylor noted that this was not as easy to detect in a standard medical exam from a regular doctor. "I must say . . . this is a little subtle, and physicians not used to examining eyes closely might conceivably overlook that." Taylor added, "A small portion had gait ataxia, five, one of whom could not stand alone." Doctors had also checked sperm levels, and in fourteen cases all were "abnormal," with the men being "relatively sterile."[72] At the hearings, in front of cameras, the workers were shaking visibly, especially if they took a drink or held something in their hand. The sight added to the sense of fear and uncertainty. In immediate testing, their livers were enlarged and tender but fortunately did not show signs of cancer at that point. Yet the long-term concern remained. As Taylor said at the House hearings in Hopewell, we "presume that this chemical is quite capable of producing cancer in humans."[73]

Taylor's colleague, Dr. Sidney Houff, emphasized the emotional and psychological aspects related to masculinity as well.

> I think that the medical aspects as presented by Dr. Taylor are as good as you can find. I think there is one aspect that we have not considered yet. With my experience with the group of people that we saw at MCV, beyond just the medical problem that they had, they have some severe emotional problems resulting from this terrible exposure. To be sterile for any man, I think, is a very emotionally devastating thing. Even if their sperm does come back, the question arises whether the sperm are abnormal or not. We cannot answer that. Another problem is the fear of cancer. How do you plan for children and how do you plan for life with that fear in the back of your mind? I think a third thing that concerns them all is their families. They are worried about whether their children are sterile. They are worried about whether their children are going to develop cancer.[74]

But there was some hope, as doctors figured out a way to remove Kepone from the workers' bodies. Taylor described how Kepone exhibited "an enterohepatic circulation": "That is[,] the chemical is partitioned into the liver, excreted into the bowel and reabsorbed by the small intestine, completing the circuit. Unfortunately, the chemical does not get out that way. If this is true a variety of means are being devised to try to capture the chemical either in the small intestine or to get it out of the blood. These are ongoing."[75] Taylor would recall in 2017 that he was optimistic that the workers would get better. He was interviewed for the *60 Minutes* episode described earlier but did not appear in the final cut. "Dan Rather interviewed me. And he wanted to know what was going to happen to these guys. Well, I looked up the other organochlorine insecticides, what happened to the other people. And there have been several fairly large epidemics. And they all got better. I told Rather, I said, "I think they'll get better." This was fairly early, and I wasn't sure of it. But I was fairly positive. And he said, "You're sure, or you hope?" I said, "I'm pretty sure." That particular part, I'm not sure that got into what was actually broadcast."[76]

Rather's reputation in Hopewell isn't positive. Many suspect that he was looking for a conclusion before weighing the facts. Taylor recalled in an earlier interview, "[Rather said] he didn't believe me and that I didn't know what I was talking about." The story around Hopewell remains that Rather had to redo a scene in Hopewell because a car (sometimes cars) went by, and people gave him the finger.[77] Still, the story punctuated the bad press covering the city. Some looked for dark humor in the situation. The Silver Belle restaurant in Richmond created a new cocktail, the Kepone Antidote, a fruity concoction served in a skull-shaped mug with smoke bubbling up out of it.[78] Meanwhile, a bumper sticker appeared on vehicles in Hopewell with the phrase "Kepone Truckin," a play on the popular phrase "Keep on Truckin." Although they insisted it had nothing to do with Kepone, at some point in the summer of 1976, Hopewell officials removed the road signs reading "Welcome to Hopewell: Chemical Capital of the South."[79]

In terms of removing Kepone from the workers' bodies, Taylor was correct. Led by Philip Guzelian, he and other doctors came upon the idea of using an cholesterol drug, cholestyramine, to treat the men. It was a novel idea. At the time, there were no documented poisonings of humans with chlordecone, no critically evaluated methods of measuring it in biological material. After getting positive results with rats, they obtained funding from both the National Institutes of Health and Allied Chemical to conduct trials on the workers. Over eighteen months of observation, doctors found that cholestyramine bound the Kepone "in the intestine to prevent reabsorption into the bloodstream," allowing it to exit the bowels more easily, from about 165 days to 80 days in blood, and 125 to 64 days in fat.[80] With quicker exit, symptoms such as tremors could diminish faster and chances of long-term dangers like cancer would be reduced.[81] Guzelian also conducted later tests and found that the workers' liver functions and sperm counts had returned to normal as well.[82] He noted how, unlike other organochlorines such as Mirex or DDT, Kepone binds more to the liver and blood than it does to fat. They also noted that Kepone behaved differently in rats and mice than it did in humans. In humans (as well as gerbils and pigs), Kepone is converted in the liver to "a reduced form, chlordecone alcohol," whereas no such transformation occurs in rats or mice. Yet it was reported: "It is unknown whether conversion to chlordecone alcohol would have any effect on chlordecone's carcinogenicity."[83]

It is not clear if all thirty-two men who were part of the testing received cholestyramine. A 1980 *Hopewell News* article noted that of the thirty-two men in the hospital, twenty-two received the treatment; ten rejected it.[84] In peer-reviewed publications, Guzelian stated that "all patients" were given cholestyramine once the trial had ended, but this likely meant of all of the twenty-two who consented.[85] Five years after the incident, Guzelian studied twelve of the most affected workers at MCV. In summary, he found, "Kepone levels were zero. Most of them had returned to work of some sort. Two or three had fathered normal children."[86] Fortunately, the tremors, nervousness, and rapid eye movements had abated.

But some effects did linger. Bill Moyer was a supervisor at Life Science and part of the Gilbert lawsuit. He reported in 1985 that the Kepone was nearly gone from his body. Although doctors in 1976 feared he might not work again, he did. Yet he still suffered from "impaired vision, occasional memory loss and arthritis," all of which he traced to Kepone.[87] Although he was one of those hospitalized at MCV, Frank Arrigo says that he did not receive cholestyramine. It is not clear if he was among the ten men who rejected the treatment or not. As of 2019, he still had 4.3 ppm of Kepone in his blood.[88] Both he, his sister, and his brother have scleroderma, a rare autoimmune disease characterized by abnormal thickening or fibrosis of the skin, joints, and internal organs. The causes are unknown, but several environmental factors show increased risk for it, including exposure to "silica dust, vinyl chloride, epoxy resins, and other organic solvents" as well as aromatic hydrocarbons such as benzene, toluene, and xylene.[89] "But see back then we didn't know about this stuff," he says, "so my clothes got washed, and [my] sister's [and] my brother's, [and] now all of a sudden I'm thinking maybe I brought that stuff home myself. Because what I have is not hereditary. I've never heard of more than two people in a family having it, but now there's three."[90]

In 1976 Guzelian's cautiously optimistic results were in the future. Scientific study of Kepone had only begun as the lawsuits made their way through the courts. Taylor believed the workers would improve once the Kepone stopped entering their bodies, but at the time he wasn't certain about long-term effects. "We will just have to say that we do not know what the ultimate prognoses are. This is with particular relevance to the relative sterility as well as to the possibility of carcinogenesis."[91] The fear and uncertainty weighed on the workers and all those involved in their cases against Allied and Life Science that began in 1975.

Settling the Workers' Cases

U.S. District Judge Robert J. Merhige Jr. would preside over the three criminal and four major civil Kepone cases at the federal level. Merhige, of Lebanese ancestry, was born in Brooklyn, New York. He earned a basketball scholarship to High Point College in North Carolina and graduated from the University of Richmond law school in 1942. He then flew forty-eight combat missions in World War II as a radar specialist on Army Air Corps bombers. He became a seasoned trial lawyer in Richmond and entered politics by cochairing a Virginia committee to elect Lyndon Johnson in 1964. LBJ repaid the favor by appointing Merhige to the federal bench in 1967. As a judge, he was known for his integrity, humor, kindness, and not suffering fools or delays in the courtroom. It is rumored that he once ordered his own father out of the courtroom for falling asleep. It was Merhige who played a critical part in making the Eastern District of Virginia known as the Rocket Docket. "I am a runner," he once said, "not a jogger." Bill Cummings, who led the federal response against Allied in the environmental arena, remembered Merhige, who passed away in 2005, as "a bigger-than-life character," a "tough judge" with a "heart of gold." "You may not always agree with his ruling, but you know he was doing it because of what he thought was the right decision, not because he had any other inclinations."[92] In his thirty-one years on the federal bench, Merhige would preside over several important cases besides Kepone—and not shy away from his verdicts. Merhige wrote the decision that threw out the appeals of Watergate figures G. Gordon Liddy, Bernard Baker, and Eugenio Martinez. He ordered the University of Virginia to admit women in 1970 and clarified the law allowing pregnant women to keep their jobs. He oversaw the Dalkon Shield Claimants Trust that paid out some $3 billion to those harmed by an intrauterine contraceptive device (IUD). His most controversial decisions were those in 1970 and 1972 ordering busing to integrate Richmond area public schools. For these decisions, protestors paraded weekly outside his home, spat in his face, and burned down a guest cottage on his property. They even shot his dog to death after tying its legs.[93] One of his sons, Mark, who now works in development in Richmond, remembers trying to live a somewhat regular life as a child then. But his father's reality made that impossible. "Eventually it was unavoidable, when the FBI is sitting in your den every day with a tape recorder, taping every phone call that comes through, and you got federal marshals surrounding the house." "It was an interesting childhood, I'll tell you that," Mark Merhige says. "[My dad] loved the law. It was his hobby as well. He didn't golf. He didn't fish. He'd have rather been reading the law or having a discussion of the law with a bunch of lawyers or teaching the law." He had an ability, Mark recalled, "to get people sitting down at the same time and finding that common ground and finding that solution."[94]

The major legal cases involving the workers and their families lasted about a year. They were settled without a jury trial. What remains of the cases are memories of the discussions, some extant court documents, and newspaper accounts. One of the chief lawyers for the workers, Robert B. Smith III, later recalled how the Kepone case took over his work. "We virtually gave up our other practice. We lived in constant fear of being in bankruptcy." They worked eighteen-hour days going through all the documents they received. They were looking for a "seminal document that tied Allied to the knowledge of what Kepone was." "We found that document in something called the Larson Toxicological Studies," Smith said. These were the studies from the 1960s that showed Kepone to cause cancer in animal tests. The bench trial date in this case continued to be postponed as evidence and motions came in from plaintiffs and defendants. In early August 1976, Allied

Judge Robert J. Merhige Jr., who was at the center of many high-profile cases during his time on the federal bench for the Eastern District of Virginia. He called the Kepone disaster "a crime against every citizen." Courtesy of the Merhige family

asked that the workers be bussed to Duke University in Durham, North Carolina, to undergo testicular, brain, liver, and kidney biopsies. Several workers expressed concern over these tests. Legally, Smith recalled, they "had no evidence" to stop it, but they made a motion to Merhige, essentially "begging" him not send them to Duke. Merhige heard the motion on Sunday, August 8, and, as they were discussing it, "in came from the side door this fellow, long beard in a running suit. It was Dr. Philip Guzelian." He asked Smith and Taylor if he could testify, and they allowed him. Guzelian told Merhige "in essence if these people are tested under the methodology in Durham, it won't be the practice of medicine, it will be the malpractice of medicine." Dr. John Taylor also testified that the tests were unnecessary. Both doctors noted that the tests would interfere with the ongoing efforts to use cholestyramine to treat the workers. After this, Merhige sided with the workers, and the tests did not happen.[95] Smith also stated that another memo helped settle the case. It was an internal Allied document "dated sometime in the early summer months of 1975 essentially saying, 'We have a satellite in the Richmond area that is having difficulties. People are sick. What should we do?' Back came a response from the board of directors, 'continue production.' With that we took that document to Judge Merhige and our case had a speedy conclusion."[96]

On December 30, 1976, the three main civil suits filed by workers and their families, *Dale Gilbert v. Allied*, *Jerry Collins v. Allied*, and *Janice Gilbert v. Allied* were settled for undisclosed amounts, although an Allied spokesperson said in a later news report it was around $3 million total for all workers before various costs were deducted. A *Fortune* article on Allied in 1978 stated that it was a $3 million settlement, with $1 million going to the workers' lawyers. Allied covered this under a general liability insurance policy, so it did not affect the company's bottom line.[97] The workers' lawyer Edward Taylor said it was a "fair settlement." Allied issued a statement calling the settlements "favorable to the interests of this corporation and its shareholders." Protracted litigation would "serve neither the interests of all litigants nor the concerns shared by the public in Virginia for a prompt resolution of the Kepone matter."[98] In part, Allied also agreed to the $3 million settlement to avoid an even costlier litigation. Lawyers' fees alone would have been substantial. By December, Allied had already been hit with a criminal fine as well, and the company's senior leadership was ready to move on. In addition, a group of stockholders was in the process of suing Allied's board of directors, claiming the board had violated its duties over Kepone. Allied and the two railroad workers settled their case in December 1980.[99]

At a panel discussion twenty years later, Judge Merhige expressed relief that the workers did not appear to suffer long-term health consequences like cancer or sterility from their exposure. "While I was waiting for the presentence report, other suits were developing, about fifteen or sixteen of them. We thought the original group of people who had been allegedly injured were horribly injured. There were reports that their reproductive capacity had gone. In any event, the case was settled well before we realized or got the reports from the doctors that the injuries were nowhere near as bad as we had first anticipated, thank God. That was one of the happy things."[100]

Some workers were less sanguine. Arrigo remembers that he and others had to keep silent about the monetary amounts and that all workers had to agree. "They sat us all around this table, an oval table—there was about twelve of us, I guess it was—and gave each one of us a paper, and they said you all can't tell each other how much your settlement is. And if everybody doesn't take the settlement, then nobody will get it." Arrigo says, "I was awarded $64,000 [but] ended up with $22,000."[101] Arrigo and Fitzgerald both stated that they had to pay out of the settlement for the battery of tests done at MCV. Arrigo: "And then, as soon as I got my first check, I noticed that every test they had done on me, I had to pay for it myself. I mean, what kind of sense does that make? I volunteered for the test." "I had the same misunderstanding," Fitzgerald recalls. "I thought that it was all part of the legal machinery or maybe some well-intentioned effort of MCV to look out for us. So, that was a significant bill, when it arrived."[102] When interviewed in 2017, both lived in modest houses. It's uncertain how much others received and how they used that money. In a news report on the tenth anniversary of the disaster, Nicky Shown, who suffered some of the worst poisoning, was said to have also developed serious mental problems. "For about five or six months, I didn't know if I was coming or going." As of 1985, he hadn't worked since leaving Life Science. He squandered most of his money because of mental problems. "I had people who took advantage of me, and I didn't know any better. I bet I've wasted every bit of $50,000, if not more."[103]

Janice Denton, who was the office administrator for Life Science, offered a mixed assessment of the lawsuits from workers. Both her sons worked there as teenagers after school. Her father worked for Allied. They did not sue.

> INTERVIEWER: Was there resentment from folks in the community against the workers who did sue, that they shouldn't have done it, that they were doing this just to get the money?
>
> DENTON: Well, I believe a lot of them were. I think the ones that really did get ill, that was fine. You needed to sue, because there were a lot of expenses, and Allied was the mother company, so Life Science didn't have that money. I know that. They wouldn't have gotten a penny out of Life Science, so they had to go to Allied. I really wanted them, Allied, to fight it, go to court. Don't do these settlements, but I didn't feel Allied had done anything wrong. I felt it was more the responsibility of some of these workers who had done things that they costed on themselves.[104]

Steve Keavy also did not sue. He recalls initially being unconcerned over filing a lawsuit but now feels more supportive of those that did. "Whether these people were permanently injured, to this day, I don't know. I heard that there were reproductive issues with some of them, and if you just keep it in that context itself, how do you compensate somebody for that? You can't, and, in retrospect, no matter how much money they paid somebody that had those type issues, it wasn't enough. Did I feel like that when I was 25? I'm sure not, but through life, yeah. If, let's say, Nicky Shown never had a child because of it, he didn't get enough, no matter what they gave him. It wasn't enough. So, from that standpoint, yeah, they had to be held responsible. If you don't, it just leads to more mistakes in the future."[105]

The settlements provided one set of residues from Kepone, remaining not just in a limited set of records in the archives but also in the living memory of those who experienced it (and in Frank Arrigo's case, in his body). As these civil cases and government inquiries over poisoning bodies went forward, so too were those addressing the pollution in the water, adding another layer to the residual governance of Kepone.

CHAPTER 5

A Crime against Every Citizen

Complicating matters, investigations and legal proceedings on Kepone's environmental pollution moved forward at the same time that the workers sought restitution in court. Contamination of the waterways brought into the Kepone story a range of individuals and industries relying on the James River for survival. Inquiry into the scale and scope of the pollution revealed the roles played by managers at Allied and Life Science, Hopewell city officials, and state and federal authorities in the disaster. In the end, legal precedents were set that had important implications for corporate and government responsibility, science and public policy, and pesticide regulation. These legal decisions provided additional residues of the toxic contamination associated with Kepone.

Finding and Not Finding Kepone in the Air

Life Science accidentally came to the attention of regulators. It began not with poisoned workers or polluted water but with air. Problems at Life Science started immediately upon production of Kepone, and the inauspicious beginning foretold of trouble ahead. During the first production run on Wednesday, February 21, 1974—the same day Virginia went on a gasoline rationing plan for motorists (those with odd-numbered license plates buying on odd days, those with even numbers on even days)—a faulty seal let sulfur trioxide escape, which gave the plant the "appearance of being on fire, [with] smoke-like substances coming out of every opening." In the *Hopewell News* report on the incident, Hundtofte said the smoke was "aggravating, but not harmful" and operations continued after an hour of clearing.[1] But SO_3 is very toxic. Its vapors are poisonous if inhaled, and limited exposure can cause burns, and irritation of skin and the lungs. It is likely that the SO_3 reacted to the air and turned into sulfuric acid (H_2SO_4) as it moved along. Sulfuric acid is a contributor to acid rain.[2] The release lasted for one hour, and the fumes "drifted across" Randolph Road (Virginia State Route 10), "impeding both vehicular and pedestrian activity."[3] Other malfunctions followed on February 27, March 4, and March 5. One of the latter was followed by fumes right over the nearby Hopewell Fire Department, which contained a Virginia Air Pollution Control Board (APCB) monitor. This, court records show, "alerted field personnel of the plant's existence."[4]

Until these events occurred, Life Science was bureaucratically invisible. The March incidents changed that and provided the first opportunity to address Kepone contamination. On March 6, 1974, Jon R. Carroll became the first governmental official to visit the plant. As the Virginia air pollution control officer for that area, he toured the plant with Hundtofte and Hopewell's fire chief. Although they had filed for corporate status, neither Hundtofte nor Moore had applied for the required permit from the APCB to construct their facility. Carroll cited Life Science for failing to get the required permit for construction and had Hundtofte install what they called a bagging operation to catch the dust from the production area that dried the wet Kepone into powder. Even so, the bagging operation did not go into effect until October 1974. Meanwhile, Carroll continued to perform "visible emission" checks of the dryer into September 1974. He did not report violations of emissions, despite others like local reporter Kit Weigel who noted the regular dust clouds. In one of his reports, Carroll mentions that a police report had been filed against Life Science for dust, but production continued. He also saw workers coming into contact with Kepone inside the plant. He wasn't concerned. It "was a dusty, dirty operation . . . but then there are a lot of industries that are dusty and dirty." Carroll commented that he asked Hundtofte what the plant made and said Hundtofte replied that it was a pesticide ingredient: "He allowed me to read the process description and nowhere did it indicate it was a very toxic substance."[5] A later report from James Kenley, who was Virginia's deputy commissioner of health and head of the interagency Kepone task force set up by Governor Godwin, noted that in March 1974 the application materials submitted by Life Science for a permit from the APCB identified Kepone as the product they manufactured but did not describe is as toxic or hazardous.[6] In January 1975 a Hopewell resident phoned the APCB about dense fog reaching across to City Point, likely containing toxic SO_3. This prompted several discussions between Hundtofte, Carroll, and other officials with the APCB, but these officials took no action against Life Science as SO_3 and Kepone dust continued to pollute the surrounding area. From May through July 1975, Carroll continued to report his "discussions" with Hundtofte and APCB officials, one that included the Hopewell Fire Department and Weigel of the *Hopewell News*.[7] It is not clear from records what, if anything, transpired from these talks.

Air pollution wasn't new, or often news, in Hopewell. Kepone dust drew little concern. In 1974 the city's large industries including Allied already exceeded limits set by the Clean Air Act on pollutants listed by the EPA, although those were going down.[8] Kepone was not listed as a toxic substance, and so the APCB did not measure for it in the filters it installed to monitor air quality. While the SO_3 leaks did warrant a visit and some concern, Kepone did not rise to level of alarm. William R. Meyer, chair of the APCB, explained to the Senate subcommittee in 1976 that the data on Kepone supplied to his agency once the SO_3 leak came to its attention "gave no appraisal of its toxicity or potential hazardous nature." Regarding Kepone, he said, "Observations of particulate emissions indicated that existing controls were not adequate. The addition of a baghouse collector was recommended. This was installed in October 1974. From that time until the plant shutdown, no further excess particulate emissions were observed or made known to us." The state tested for levels of particulates in the air, not toxicity. Meyer noted, "[There] was no indication of our readings at the Hopewell building to indicate that the amount of pollution at that site had gone up. In fact, the trend since 1970 was down."[9]

This was the first chance to halt Kepone production. Another opportunity to shut down Life Science emerged over Kepone dust. Orben R. DuBose worked in the dryer area surrounded by the white powder. His supervisor, plant manager Del White, told DuBose to unclog a rotor that fed Kepone into the dryer. DuBose refused, saying he couldn't work anymore under those conditions. White fired him for insubordination. DuBose had previously been reprimanded for failing to do his job, including one incident in which White and Hundtofte both claimed DuBose was intoxicated. After being fired, DuBose filed a complaint with the federal Occupational Safety and Health Administration detailing the hazards of dusty conditions and the health concerns for himself and fellow workers. He made a written complaint and also spoke in person with an interviewer in Richmond. In October 1974, rather than inspect Life Science, OSHA's industrial hygienist responsible for the region that included Hopewell (with some thirty-seven thousand firms to monitor) reviewed DuBose's complaint and the limited material in his possession on Kepone. He determined that it did not warrant his immediate attention. Finding that Kepone did not rise to the equivalent of other toxic substances with which he was dealing, such as lead and silica, the official "put it in a pile of 80 other referrals on health matters that he had to treat in sequence," as OSHA official Morton Corn would report at the House hearings in 1976. Instead, the agency determined that DuBose was claiming discrimination for being fired as a result of filing the complaint (which was incorrect), and the agency could find no evidence of this. Other agency personnel failed to follow up on the health issues DuBose raised, and OSHA remained out of the Life Science facility until after it was shut down.[10] As Senator Allen commented to Corn: "I want to say that you were thrown off the scent mighty easily."[11]

Kepone in the Water

Toxic waste from Life Science that entered Hopewell's sewage system caused more concern. This waste fell under laws governing pollution in federal waters such as the James River. The first major law still in effect was the federal Refuse Act of 1899. The act imposed criminal liability on polluters for discharging into navigable waters or tributaries of those waters without a permit. The courts considered each discharge to be a separate event, thus opening the possibility of polluters being fined for each day of a discharge. The law placed the U.S. Army Corps of Engineers in charge of granting the permits. As environmental pressures mounted in the late 1960s, news reports and federal investigations by Congress and the General Accounting Office revealed that the Corps of Engineers had been lax in its permit-granting duties, neither encouraging companies to seek them nor actively investigating whether companies possessed them. Although the Corps of Engineers responded in 1970 by announcing a renewed effort at enforcement, President Richard Nixon went further and issued Executive Order 11574, directing the Corps of Engineers to work with the new EPA on a revived permit program, the Refuse Act Permit Program.[12] Although federal courts struck down that program in 1971, the idea of the permit program survived and became part of the federal Clean Water Act of 1972. The new permit program was the National Pollutant Discharge Elimination System (NPDES), focused on point source pollution, and it required industries (and eventually municipal sources) to abide by standards for effluent limits of pollutants.[13]

In keeping with ideas of cooperative federalism, under the NPDES the federal government set standards, and, if a state met those standards, that state managed the process of obtaining a permit. States can adopt stricter standards but not standards that are less strict. In the Senate subcommittee hearings on Kepone, EPA administrator Russell Train noted that, in terms of approving a permit, it was really the state "that pulled together the information and, in effect, approved the permit." "[The state] transmitted it to us, and if [the body seeking a permit] had met the State's requirements, we fairly automatically [approved it]." Regarding Hopewell, the city applied for the new permit in October 1973, before Life Science came into existence. As Train noted, at that point there was no mention of "industrial discharges going into the system."[14]

That changed when Life Science started up. In 1974 industrial wastes from major plants like Allied went into the waterways directly, not through the city sewer system. These factories installed their own equipment to treat waste before dumping. At this point, these industries were involved with Hopewell and the government to build a new regional sewage treatment facility that could handle industrial waste. This would not come online for a few more years. How was it that the city allowed Life Science to discharge its toxic waste directly into the city's sewer system, and did the city notify the state or EPA of the change to its NPDES permit? It's not clear exactly how the approval process unfolded between Life Science and the City of Hopewell. According to former city manager Arthur Lane, Life Science wrote to the city in November 1973 requesting to discharge directly into the sewer system. Given that the site on Randolph Road did not border a waterway, it seems that without this approval the plant could not run. This letter listed Kepone as one of the components to be discharged. Lane approached Clifton L. Jones, Hopewell's liaison engineer for the new sewage treatment plant and also in charge of working directly with Life Science. Jones had been the plant manager for Allied's agricultural production that made Kepone and other products before Hundtofte took over that job. Lane understood that Jones knew about Kepone, having worked at Allied. Yet Lane said it appeared to him that Jones "was not familiar with Kepone either." As far as Lane was aware, Jones contacted Hundtofte and Moore, and they assured Jones that "Kepone was safe and could be discharged into the city sewerage system without any harmful effects." Based on these assurances, Lane granted the request, with the conditions that the city's treatment system not be damaged, that the city could require monitoring and more information from Life Science, and that Life Science inform the city of any changes to its effluent. With this letter in hand, Life Science moved forward on its contract with Allied.[15] Moore and Hundtofte obtained permission from the City of Hopewell to discharge "approximately 500 gallons of waste water per hour" into the city's sewer system, claiming the waste would contain "three parts per million of Kepone" as well, an amount that would not harm the city's treatment system.[16] Accordingly, Life Science became the only industry in Hopewell allowed to discharge directly into the municipal sewer system.[17]

According to William Moore, Jones assured him that Life Science did not need to inform the State Water Control Board (SWCB) since Life Science wastes were going to the city's treatment facility, not directly into the river. Under the NPDES permit system, permission to discharge wastes went through the city, not the state. The SWCB chief administrative officer Eugene Jensen confirmed this during Senate testimony.[18] Under this arrangement, the SWCB—the agency responsible for assuring water quality—had no knowledge of Life Science or its agreement with Hopewell. The city also failed to notify the EPA of these changes to its original permit filed in 1973. The last update the EPA possessed from Hopewell was in October 1974, and that did not reference industrial wastes being discharged into the city's sewer system.[19] All of these would prove to be fateful decisions.

First indications of poison in the water emerged when Kepone waste shut down critical operations of Hopewell's Sewage Treatment Plant (HSTP). The system in place at the time was a primary treatment system that divided water and solid waste. Chemicals treated the water before the effluent entered Bailey Creek. Solids were separated into sludge and treated with bacteria in a digester to reduce the overall amount and remove organic pollutants. The sludge was then trucked to the city landfill and buried.[20] The first waste from Life Science overwhelmed the system, killing the bacteria and requiring officials at the HSTP to construct 100x150-foot containment lagoons to hold thousands of gallons of the "bugless sludge."[21] At first, public works employees weren't sure what was killing the bacteria. But tracing in reverse the acrid smell of effluent that led into the treatment plant, going from sewer hole to sewer hole, they found their culprit: Life Science. When this happened, city officials failed to inform either the state or the EPA, choosing to work with Life Science to address the issue. In May 1974, city officials insisted that Life Science install a "primary sedimentation tank" to remove some Kepone solids in the wastewater from their plant. But during the summer of 1974, serious problems with operations continued, and William Havens, who operated the Hopewell Sewage Treatment Plant, "continually noticed the dark colored water with white, flocculent particles, characteristic of Kepone." In July the city had Life Science install "a centrifuge and cloth filters" to aid in removing Kepone. Hundtofte insisted that the filters "reduced Kepone emission into the sewage system to 'virtually zero.'"[22] These steps did remove much of the Kepone from the wastewater, but the filters were often blocked. While they were being cleaned, untreated Kepone waste went into the sewer system. There were also significant and regular "spills, overflows, and leaks" that escaped containment. Even with these precautions, the HSTP was not designed to remove Kepone or the component materials used to make it. Throughout this process, the City of Hopewell never disclosed to the SWCB the issue with the digesters, and, despite the efforts to contain it, Life Science continued to dump "between fifteen and twenty thousand gallons per day of Kepone contaminated waste water down the Hopewell sewer."[23] Moore and Hundtofte admitted that they failed to test to make sure they were within the limit of 3 ppm. Only "eyeball tests" were made—if the liquid waste was clear, they assumed it was okay.[24]

It wasn't until October 1974—seven months after first encountering the Kepone waste—that the SWCB became aware of the issue. Like with air pollution, it seems to have happened by accident. One version of events holds that on October 3, 1974, Havens of the HSTP phoned the SWCB to inform the board of the problems. Havens also asked the SWCB "not to disclose that it was he who had reported the problems, for fear that action might be taken against him by Hopewell officials." Havens also "approached C. L. Jones several times prior to October 1974, about the Kepone discharge and the problems he was having." Havens claims that Jones insisted that "Kepone could not be affecting his plant performance."[25]

Another version, with greater documentation, centers on SWCB staff engineer

John Reeves, who was fresh out of college. In March 1974—just as Life Science began its operations, Reeves joined the Piedmont office of the SWCB as a staff engineer. He was assigned to work with Hopewell on the new treatment facility. In September 1974 Reeves arranged a meeting with Jones regarding the building of the new regional sewer plant in Hopewell. At the meeting, Jones mentioned the problems with Life Science. As Reeves told the Senate subcommittee in 1976: "I will bring up for the record that when I was told of [Life Science] informally by Mr. Jones, who worked for the city of Hopewell, I did not visit the plant." Reeves told his boss, who, according to Reeves, said that "since Mr. Jones was a widely respected engineer and had told us he would fix the problem; that we would give him 2 weeks, and then we were going to do something." In this intervening two weeks during which the state, without an investigation, trusted Hopewell to fix the issue, the phone call from Havens may have come in. In early October, possibly October 3, another engineer with the SWCB, W. W. Floyd, visited Life Science and said that "something should be done right away." Floyd discovered that the digesters at the Hopewell Sewage Treatment Plant were "essentially inoperative," and it appears that the board directed the city to cease dumping its "illegal sludge."[26] Reeves was assigned to follow up. "On October 7, I visited the sewage treatment plant, which is under our water board's control. I talked to the operator. Then I went to talk to Life Science. This meeting was on the 7th. I believe it was in the afternoon, at which time I talked to Mr. Moore and Mr. Hundtofte." Reeves commented, "[They] reported some very sketchy information on what they did, and agreed with me at the time to have Allied Chemical Corp. do testing on the sludge at the sewage treatment plant and on their discharge of water." They used Allied because the company had the facilities and because of the relationship with Life Science. The first sample held 3,200 parts per billion (ppb) of Kepone; the second had 68,000 parts per billion. Put into perspective, later recommendations on safe Kepone amounts in effluent would be .5 parts per billion. This prompted the first meeting of the SWCB with the city as well as Hundtofte and Moore, on October 31, to determine the extent of the problem and consider a solution.[27]

According to a memorandum written by Reeves, at this October 31, 1974, meeting, Hundtofte acknowledged the problem and noted that Life Science had installed filters and settling tanks to address Kepone in the wastewater. Another issue that came up was the cost to Life Science for these measures. Reeves wrote that Hundtofte stated these cost Life Science $25,000, "a heavy expense for a small company." Those in attendance at the meeting likely did not know that Allied would pay those costs—it doesn't appear that Hundtofte or Moore let anyone know. Allied's Agricultural Division president G. C. Matthiesen would comment in the 1976 Senate subcommittee hearings, "[Life Science was] fully aware of environmental concerns, having requested and received from us assurances that necessary environmental capital expenditures would be recovered as a surcharge on each pound delivered." According to the contract between the two companies: "If Allied Chemical agrees that the plant can be modified, L.S.P. shall effect such modifications and Allied Chemical shall pay for same over the course of the twelve-month period following installation of such modifications." Moore was aware of this too. In a letter to Allied on February 25, 1975, he noted, "[To] cover capital expenditures for pollution control requirements, as provided in our contract, we propose to continue the $.23 per pound of Kepone delivered, capital repayment charge."[28]

David S. Bailey played a key role in this story. He began as a field biologist with the SWCB and then became the agency's director of ecological studies. He gathered and interpreted literature on the toxicity of Kepone, and those at the October 31 meeting read his recommendations. Bailey advised a concentration of no more than .4 ppb in effluent released into the James River. According to Reeves, "Mr. Moore strongly questioned the toxicity data and interpretation and proposed to resolve with Mr. Bailey questions that he felt additional information would support. I stated that this toxicity position is based on the best available information and would have to be proven incorrect by the company." Jones also insisted that Life Science "prove to the City soon that they have reliable, continuous monitoring and a discharge that is non-toxic" or stop discharging to the city's sewers. What Jones meant by "soon" is unknown, as Life Science had been discharging waste for seven months at that point.[29] No formal plan came from the Halloween meeting, and none in authority seemed to suggest Life Science shut down production.[30] They agreed to meet again on November 21.

When the *Hopewell News* covered the topic in November 1974, Jones showed that he seemed to know a good deal about Kepone but tried to dismiss the problem. Lane defended Jones, saying that it was "absurd for anyone to assume Mr. Jones was aware of any potential dangers and did not tell us about them." Lane trusted Jones, and Jones certainly trusted his former fellow Allied workers Hundtofte and Moore. He told the *Hopewell News* that Kepone was not biodegradable and that it "decomposes slowly." "It is lethal to aquatic life in large enough amounts . . . and is harmful to humans and other warm-blooded animals because it accumulates in the body like DDT." He dismissed concerns, though, saying that some Kepone was probably discharged into Bailey Creek before the problem was found but that it "would not be enough to be harmful."[31] Later, when questioned about the city's role in the Kepone disaster, Lane claimed ignorance, made it appear that the state was dragging its heels, and blamed Life Science. "We didn't know anything about what we were dealing with and each time Life Science kept giving us assurances that the discharge was not harmful." Until October 1974, he said, the city had gone from one state agency to another trying to find out the problem and what could be done about it. "Because Kepone was an unknown, no one seemed to have any answers."[32]

At the 1976 Senate subcommittee hearings, Eugene Jensen, executive secretary of the SWCB, provided a summary of the November 21 meeting in 1974. There's no indication that Life Science presented a plan or showed progress on reducing effluent. Moore tried to offer some information on his and Hundtofte's efforts and what they would consider a safe level of Kepone in the company's effluent. According to Jensen, "My staff stated that Moore's information was not useful in defining the staff's determination." Fred Hughes, Hopewell's public works director, issued a public comment on the situation after this inconclusive November 21 meeting. Later, in the Senate hearings, John Reeves would dispute key parts of Hughes's statement. Hughes argued that he and his staff were ready to address a problem in Hopewell's treatment plant, but were "caught in the middle of a bureaucratic mish mash." He was optimistic that the state Health Department would soon approve a plan to dispose of the toxic sludge.[33] Reeves bristled at Hughes's characterization. "Mish Mash? The SWCB staff and SHD [State Health Department] have common goals of protecting human health and the environment from such toxic materials as Kepone." Reeves accepted Hughes's frustration at determining a safe level of Kepone. "However," he said, "our technical analysis of this pesticide concluded that the environment should have less than 0.4 ppb Kepone, while the Health Department says 0.00 ppb to the environment and to the STP [Sewage Treatment Plant]. Hopewell does need direction." Reeves also corrected a statement made by Hundtofte, who claimed that Life Science had installed a filter system that had reduced the Kepone emission to "virtually zero." Reeves fired back: "'Virtually zero' is more like 1–2 ppm, and their leaks and spills add much more." Hughes also stated that the city had stopped testing for Kepone entering its sewage plant recently because Life Science had not been operating at full capacity in several weeks. The implication was that the city did test earlier. Reeves noted wryly: "Three weeks is more like seven months."[34]

Reeves may have been the most vocal in demanding more action against Life Science. But as a junior engineer he lacked authority to act on his own. Still, shutting down the plant never became a serious possibility through early 1975. Senator Leahy asked Reeves directly: "Did there come a time when you recommended it be closed down or at least [that] the process that they were following be stopped?" Reeves replied: "I believe I recommended to my superiors that action be taken and pretreatment be required, and that if something was not done pretty soon, we ought to pool all the forces that we could and do something to them. I have not any memo, to my recollection, saying that they should be shut down before some other action was done."[35] Life Science's Moore indicated in the hearings that at one meeting the city attorney for Hopewell, Torsten Peterson, suggested shutting down Life Science until they could get the effluent under control. "Nobody ever said anything about shutting us down, save Mr. Peterson." Moore recalled that Peterson said, "Why do we not just shut you down and when we are convinced that the effluent is safely under control then you can start up again." It was a simple invocation of the precautionary principle that went nowhere. "Somebody disagreed with that," Moore recalled, but he could not remember who it was.[36]

In an exchange with Leahy, Virginia's assistant attorney general David Evans defended the decision not to shut down Life Science in late 1974 once it was clear the company was polluting the river and damaged one of Hopewell's digesters. There was said to have been a meeting in March or April 1975 with Reeves, Raymond Bowles, who was the director of the Bureau of Enforcement for the SWCB, and others. Evans said,

> Progress was being made, we believed, to reduce the quantity of Kepone going into State waters.
>
> I asked what technical data was available. They told me a determination had been made based upon the available data that 0.5 parts per billion could be safely discharged into the environment.
>
> They further told me the company was willing to make the necessary expenditure to achieve this limitation. My advice to them at that time was that if it was felt that that was a safe limitation, and that the company was willing to make the expenditure to do that, and it was believed that those pretreatment facilities would achieve that limitation, then there was absolutely no basis whatsoever for closing down the plant.
>
> I would still abide by that opinion at this time.

Leahy wondered, "Was there any suggestion made by either Mr. Reeves or Mr. Bowles at that time of violations of Virginia law?" Evans said no. "We were approaching it in a manner of trying to reduce the amount of Kepone going into State waters, and it was felt that at that time that definite progress was being made."[37]

Eugene Jensen sought to explain this to Senate investigators by arguing ignorance and a lack of regulatory authority: "Our problem at that time was finding out the nature of the material we were working with—was it essentially something similar to Mirex, a registered pesticide, a relatively nontoxic chemical? In fact, as it turned out, it was a toxic one. That information was hard to come by. We had many conferences trying to decide exactly how much authority we might have."[38] Perhaps, legally, since the amended NPDES permit with Kepone limits was not slated to go into effect until June, the State Water Control Board believed that they could not close Life Science down. But Jensen's claim about Mirex is interesting and more troubling. It's not clear why he would characterize Mirex as "relatively nontoxic." Perhaps he meant in acute exposure, where its LD50 rating is moderately toxic. Yet long-term exposure is toxic. Kepone and Mirex are closely related, and reports on Kepone by 1974 mentioned Mirex, along with DDT as a close relative. Dr. Chou saw this when he sought information to help Dale Gilbert. Moreover, Hooker Chemical and Allied—players in the Kepone drama—both developed pesticide formulations with Mirex from the early 1960s through the early 1970s. Data on the negative effects of Mirex was significant by 1974. The EPA had issued an order to cancel all Mirex registrations in 1971 and banned its use in 1977.[39]

It seems that during November 1974, members of the SWCB contacted the EPA to get more information on the toxicity levels of Kepone and what levels could be considered safe for discharge both to the plant and to the river. In a phone call to the SWCB, EPA staff recommended a limit of .4 ppb, which conformed with David Bailey's recommendation. Then at a December 1974 meeting—some nine months after first realizing the problem—the SWCB, in coordination with the state Health Department, came up with a set of requirements for Life Science but kept the plant open. At some point the state agreed to force Life Science to achieve a 200 ppb limit (later changed to 100 ppb), which would result in an acceptable .5 ppb at the treatment plant, slightly above the earlier one of .4 ppb. Since the plant's digesters could not remove Kepone, the plan called for the plant to dilute the Life Science effluent with other wastewater to achieve the acceptable limit for discharge into the water. Life Science was to collect and analyze its effluent. Moore and Hundtofte said they expected to meet that requirement but gave no specific date. The state also sent notice to the EPA that it would amend Hopewell's NPDES to allow for the new limits on Kepone in the discharge to the river. The process was not immediate and took some three months to get final approval, as it needed to go through public notice and have it cleared through the EPA. The state gave Life Science until June 1975 to meet the new requirements—fifteen months after first discovering the problem and with the company continuing to emit waste into the sewer system.[40] The SWCB continued to notify Life Science and Hopewell that they were exceeding the agreed-upon limits. The new NPDES permit that went into effect in June required Hopewell to sample for Kepone three times per month. But, according to the state attorney general, the city only monitored once from June through July. Also, Hopewell had been compelled to clean out and repair the digesters by June 1, 1975, but that date passed without action.[41] Neither the EPA nor the SWCB knew how much Kepone went into Bailey Creek and then the James River.

As the agencies gave Life Science more time to address the waste of which they were aware and trusted Hundtofte and Moore to be open with them, Hundtofte continued to supervise surreptitious dumping of Kepone waste. At times, the LSP facility produced excessive wastewater in addition to what came from the day-to-day production. This waste likely contained excessive amounts of Kepone above the limits set by the SWCB or contained HCP, one of the chemicals used in making Kepone but not authorized for separate discharge. One method Hundtofte devised to remove this waste from the Life Science facility involved the pickup truck system Frank Arrigo recalled. Another came in October 1974 when Hundtofte hired two septic tank operators to dump wastewater from the top of Hopewell's landfill, down the slope and into Bailey Creek, which emptied into the James River. It's not clear how many times this occurred, but at some point the landfill stopped allowing the septic trucks to dump their Kepone waste. Life Science workers even took excessive wastewater and dumped it into the sewer drains on the Life Science property, bypassing filters installed to capture the Kepone suspended in the water. Some workers, like Steve Keavy, were directed to dump waste into city sewer drains nearby. Keavy also drove to a sewer pump station blocks from Life Science to dump barrels of chlorine into the drain there ("maybe to cover up a discharge," he later surmised).[42] When chlorine became too expensive, they used Tide detergent to mask the smell to avoid detection at the sewage treatment plant.[43] Once in Bailey Creek or other waterways in Hopewell, the waste flowed into the James River, which emptied into the Chesapeake Bay. At no time in the House or Senate hearings of January 1976, nor earlier when meeting with city and state officials, did Hundtofte or Moore divulge any of this information. The sludge, with 80 ppm of Kepone, caught a lot of the Kepone, but not all. The amounts in the waterways were not fully known until the full sampling took place only after Life Science had been shut down.

In their tactic of jawboning Life Science to get compliance, Reeves, Jensen, and others from the SWCB failed in this case. It stands in contrast to how the SWCB had been acting in other environmental areas. Beginning with the appointment of Norman Cole under Governor Linwood Holton, the agency had been involved across the state in pressuring counties and cities to upgrade their sewage plants per the new NPDES system and regulations stemming from the federal Water Pollution Control Act. Hopewell was already doing this. In addition, the amount of environmental litigation in which the state attorney general's office engaged increased steadily as the environmental decade of the 1970s continued. As Jim Ryan, deputy attorney general from 1976 to 1980, later recalled, Cole and the board "picked up the cudgel" and became part of the "green revolution" to become active in enforcement. Even so, a 1976 report by the Joint Legislative Audit and Review Committee of the Virginia General Assembly criticized the SWCB for failing to address water shortages in the state and serious shortcomings in controlling pollution.[44] Meanwhile, Life Science continued to pollute and poison its workers.[45]

Only after the Health Department order forced Life Science to close in the summer of 1975 did the SWCB seek restitution. In a February 1976 memo summarizing the situation for Governor Godwin and laying out the state's case against Life Science, the office of the attorney general tried to lay blame on the EPA. The SWCB, the memo claimed, "examined the limited material available on Kepone and its effects on the environment. Between November 1974, and April 1975, the [Water Control Board] Staff repeatedly requested technical assistance from the U.S. Environmental Protection Agency (EPA) in this regard; however, none was provided."[46] If by technical assistance the memo meant research on Kepone's effects on marine life in the environment, there were, of course, no studies done. The studies that were available had been done by Allied in the 1960s to test toxicity in animals, and potentially in humans, with a focus on cancer. But EPA officials did gather information that was available and communicate this to the SWCB. In records provided to the Senate subcommittee in January 1976, one Ernest Kaeufer, an engineer in the Municipal Permit Programs Branch of the EPA's Enforcement Division, had prepared information and communicated this to Al Anthony in the Bureau of Applied Technology for the SWCB. Anthony, Reeves's memos show, attended several meetings with Hopewell regarding Kepone, and he possessed information on Kepone's toxicity on fire ants, birds, and fish. Also, at least some, perhaps all, of David Bailey's data came from the EPA as well.[47] Yet the EPA was not entirely blameless. The agency's actions were constrained by how the relevant regulations covering pesticides were written, concerns with federalism in allowing Virginia leeway in solving the issue, and a narrow view of individual and group responsibilities. The EPA staff limited its focus to finding data on Kepone's toxicity insofar as it applied to Virginia's request for setting limits on effluent. They could not inspect the Life Science plant, even if they thought for some reason to do so—the agency lacked the authority. Nor did those who were working on the problem suggest to another federal agency, OSHA especially, to do so.

Senator Allen put this point to Joseph Galda, chief of the Municipal Permits Branch of the EPA: "But you did not know whether what was going on in the plant was an operation as to solids, or dust, or fluids. You just saw the presence there in the sewage?" Galda said: "That was our primary responsibility, Senator. There are so many industries in our region; we have one man per State, and our curiosity was somewhat limited by the amount of resources." Galda did not see a connection between the effluent and anything involving conditions inside the plant. "The industry could be a perfectly clean operation, but just have a bad batch and just dump it down the sewer, perhaps. So I just didn't see any connection, or see any need for any curiosity to go up and find out how they make Kepone."[48] In this case, as Galda admitted, it was a wrong assumption. David Butler, a consumer safety officer from the Region III office of the EPA, visited Life Science on

March 7, 1975, for an inspection related to Kepone's possible registration as a pesticide, not to inspect the plant. It seems Butler did not observe anything he felt warranted notification of another agency. He met with Hundtofte in the office and did not venture into the processing areas. Butler's visit was the prompt that did eventually have the EPA classify Kepone as a pesticide and issue a stop sale order, but that was in August 1975 after reports of poisoned workers had hit the news and Life Science had shut down.[49]

Events in May and June 1975 provided additional opportunities for regulators to inspect Life Science and perhaps shut down the facilities. This time it was a different agency, Virginia's Bureau of Industrial Hygiene, charged with protecting the health of industrial workers. On May 27, one of the workers at the HSTP was overcome from fumes from HCP, a hazardous chemical used to make Kepone. Sewage treatment staff called the Industrial Hygiene office, which informed them of its toxic nature. The bureau's role was to protect worker health, yet—like other state-level regulators—it "did not know of the existence of Life Science Products" until the worker was exposed. No one from the bureau had ever inspected the plant. This became a point of contention during subsequent investigations. Virginia authorities argued that then-current law prevented the Bureau of Industrial Hygiene from demanding a visit inside Life Science, claiming that they were "pre-empted from enforcement" of industrial hygiene standards by OSHA provisions. With OSHA's creation in 1970, states needed an approved safety and health enforcement program that at least met the federal standard. In states without an approved plan, like Virginia prior to 1976, OSHA preempted state authority on safety and health enforcement where federal standards existed. Given this, according to Oscar Adams, who oversaw the Bureau of Industrial Hygiene, his agency had to "seek concurrence of the owners before making an inspection."[50] Yet OSHA official Neil Ewing dismissed this claim, arguing that Virginia could have enforced standards on Kepone "because the federal government had no standards listed on Kepone." "The state," he said, "has blamed any inaction on their part on the preemption regardless of whether it's valid or not."[51]

The May 1975 incident with the treatment plant worker exposed to HCP finally prompted a visit. On June 6, a group from the City of Hopewell, SWCB, and Industrial Hygiene met with Hundtofte and plant manager H. D. Howard. The group did not go inside the plant but saw the HCP-saturated ground near the holding tanks outside. At that point, Bureau of Industrial Hygiene director Bryce Schofield and another inspector "realized a survey of Life Science seemed to be called for, considering the toxic nature of what was being handled." He tried to arrange an inspection date, but Life Science broke the appointments "because of equipment failures." They were finally scheduled to come, but Robert Jackson's intervention closed LSP before that could happen.[52]

Federal Indictments:
Frozen Seafood and the Smoking Gun

Life Science remained in the crosshairs of regulators and workers once the poisoning and contamination of the environment became public knowledge in the summer of 1975. Yet slowly over the fall and then with growing intensity beginning in 1976, attention turned to Hopewell's chemical behemoth, Allied. Allied's initial public response was to maintain its reputation as a good corporate citizen, while distancing itself from Life Science. As one of its chief lawyers, Bob Sand, commented later: "Perhaps especially because Allied Chemical felt it was a leading good citizen, we were unprepared to cope with this type of problem." Sand noted that Allied chairman John T. Connor, a devout Catholic, had been Lyndon Johnson's secretary of commerce and resigned in 1967 over his role in economic policy and the Vietnam War. "Good guys like us shouldn't have problems."[53] John Connor himself later said: "When I first met with the Governor about the Kepone problem, I pledged that our company would do the right thing."[54] Godwin also insisted on it. Connor added: "We at least had a moral responsibility to resolve the damages and to help the people who had been injured." He took this position against the advice of the company's lawyers, who "felt that his generosity would be inevitably misconstrued as acceptance of liability."[55]

Following through from Connor's commitment, Allied took over the cleanup, since neither the state nor Life Science had the money or the staff. Allied spent $394,000 on removing the plant's machinery and placing it in the Kepone graveyard at the old Hopewell sewage plant. The company also funded studies at MCV to find a cure for the poisoned workers. As G. C. Matthiessen stated to the Senate subcommittee in January 1976: "We responded promptly and affirmatively as a good corporate citizen, not because of any legal responsibility." But he was quick to distance Allied from Life Science. "We do not wish our cooperation . . . to be misinterpreted as a concession that we are to blame or that we are legally responsible for the activities of persons not under our control." Life Science was "an independent concern with all the expertise, knowledge, and access to resources to process and handle Kepone as safely as Allied Chemical had itself for a quarter of a century."[56] Indeed, it seemed that, as of January 1976, Allied might emerge unscathed, despite the lawsuits filed by workers. But the federal indictments handed down in May 1976 sullied the "good corporate citizen" image, as they uncovered Allied's role in polluting the marine environment and called into question the relationship between Allied and Life Science as separate entities. Further pressure on Allied came from lawsuits filed by groups representing the seafood industry.

Eventually the Kepone affair made its way to the Justice Department's Bill Cummings, who would lead federal prosecutions against Allied. Cummings was born in New York but moved to Newport News, Virginia, as a young child. After college at Randolph-Macon, he graduated in 1964 from law school at the University of Virginia and went to work with Tolbert, Lewis, Fitzgerald, a northern Virginia law firm with connections to the Republican Party and the Justice Department's Eastern District of Virginia. Cummings had been "very active with the group" to get Linwood Holton elected as Virginia's first Republican governor since Reconstruction. In 1975, with Gerald Ford as president, after the FBI declared the first nominee for U.S. attorney for the Eastern District incompetent, Cummings got the appointment. As Cummings remembered much later, it was "very unusual" in those days having what is usually a civil matter of environmental damage turn into a criminal matter and come before a U.S. attorney.[57] Even more unusual was that in December 1975 the EPA made a direct referral of the Kepone case to Cummings's Eastern District office, as opposed to the larger Department of Justice to review it and make a prosecution at its discretion.

Dan Snyder of the EPA said at the time, "[Available evidence] indicates there were serious, knowing and repeated violations of federal water pollution laws."[58] Cummings would later comment in 1995, "I think they were just so concerned, they wanted to get something going, and they went directly to us. We assembled a team of prosecutors to work on the case because of having seen what the papers had generated about the incident and the disaster with the employees at Life Science, so we knew it was a big matter."[59]

In January 1976, a twenty-one-member grand jury was impaneled in Richmond. Over four months they heard testimony from fifty witnesses and examined over five hundred documents. Twenty years later Cummings recalled, "[We] were frustrated that Life Science, the one who had caused the damage in the great magnitude, was a defunct company by that time. It was [shut down]. It was bankrupt. They had only the single product that they were working on. We were trying to find deeper pockets to try and make more of an impact, and so we went looking for Allied Chemical."[60] In doing so, they came upon two critical pieces of evidence: frozen seafood and a "smoking gun" memo. David Bailey found some of the frozen fish. After working on setting limits of Kepone for the SWCB, he went into the Water Control Board's frozen-fish "library" seeking samples of fish caught in the James in the early 1970s, before Life Science got the Kepone contract. "Sure enough, they had heavy concentrations of Kepone ... and [these fish were caught] before anybody even knew Kepone existed."[61] More evidence came from William Hargis, director of the Virginia Institute of Marine Science (VIMS). In late January 1976, in front of the Virginia Senate Agricultural Committee and the Virginia Marine Resources Commission, Hargis reported that oysters with .26 parts per million of Kepone had been taken from the James four years before and kept frozen for future research and that frozen fish also from 1972 showed Kepone levels similar to those tested in 1975. In addition, Robert Jackson reported that air samples from monitoring stations in Hopewell showed Kepone as early as February 1974.[62]

Cummings came upon a key 1972 memo written by then–Allied Hopewell manager Virgil Hundtofte. According to Cummings, as his lawyers went through the grand jury documents, "We discovered an internal memo authored by Mr. Hundtofte that really got us upset, because he recognized in this memo that the Corps of Engineers' regulations required them to report any discharge of a toxic substance. There were memos back and forth saying, 'Well, the thing to do is to do nothing.' They wrote this in a memo. The way to respond to the regulations is to do nothing. So we thought we had the smoking gun."[63]

The context for those memos is important. It was during the period when federal permit regulations on effluents shifted from the Refuse Act Permit Program (RAPP) to the National Pollutant Discharge Elimination System (NPDES). At the same time, Hopewell was proceeding with its new regional plant that would handle the industrial waste from Allied and other industries. In the middle of these developments, Allied also shifted the managerial responsibilities for Kepone production at the Semiworks between 1966 and 1973. Initially it fell under the Agricultural Division's research area, where William Moore held ultimate responsibility for Kepone from 1966 to 1969. Allied then moved responsibility to the manufacturing area within Agriculture, during which time both C. L. Jones (who would later become Hopewell's liaison engineer for the sewage treatment plant) and then Hundtofte supervised Kepone production. Then in 1972 Allied executives decided to shift the Agricultural Division to Houston and place Kepone production in Hopewell under the authority of the Plastics Division. That's when Kepone shared equipment with THEIC and TAIC; at this point, Hundtofte resigned rather than move to Houston.

Federal prosecutors assembled clear evidence of violations of both the Refuse Act and the 1972 amendments to the Water Pollution Control Act (usually called the Clean Water Act). The Refuse Act, part of the Rivers and Harbors Act of 1899, as noted, makes it a federal crime to discharge refuse into navigable waters or tributaries of those waters of the United States without a permit. Prior to the case against Allied, federal prosecutors had been active in using what was an obscure piece of legislation to combat pollution. Courts had been reading the word "refuse" more openly, to include substances that "did not physically impede navigation."[64] Regarding the amendments of 1972, this legislation added standards of water quality as part of the permit process, something lacking in the Refuse Act. The NPDES emerged as part of this legislation. Under both laws, the corporations and the officers of corporations were liable, as was the NPDES permit holder, in this case, Hopewell.

It is not clear whether Allied possessed Refuse Act permits issued by the Army Corps of Engineers before 1970, although its high-level executives insisted that they did. With or without the permit, local leaders at Allied in Hopewell put Kepone, THEIC, and TAIC into Gravelly Run, perhaps untreated, before 1970. Then, at two critical moments in 1971 (before RAPP was struck down) and then 1972, executives in Hopewell had to decide how to address the new, more forceful permit requirements. In charge of responding was none other than Virgil Hundtofte. In the 1971 form required by the EPA, Hundtofte made no mention of Kepone, THEIC, or TAIC discharges. He and other executives deliberately left out the information, concerned that they would be forced to install pollution control equipment or suspend production if they acknowledged it. They also identified the wastes as "temporary" to avoid further action. This indicates that wastes prior to this may not have been treated, or, if they were, then the new requirements demanded more stringent removal of chemicals. Hundtofte and others working with him knew that the City of Hopewell was planning to build a new regional treatment facility, then set to open in 1975, and they hoped to buy time until then.[65] They also worried that mentioning Kepone might require new equipment that would cost some $700,000 or mean suspending production. Hundtofte wrote in the memo that it "was felt that this effluent might go unnoticed by the EPA until we tied into the R.W.T.P. [Regional Water Treatment Plant] or, at worst, interim treatment would not be required."[66] It appears that Allied's top management outside Hopewell did not know of the subterfuge. Maybe that explains some of the confusion and mixed messages on Allied's corporate culpability that came from Allied spokesman John Flint. He stated publicly in January 1976, on the eve of the Senate subcommittee hearings, that Allied's discharges in the early 1960s and 1970s were done "under permits from state and federal regulatory agencies." He said that Kepone was not listed specifically on the permits but that "chlorinated hydrocarbons was listed." Allied was to treat its waste with filters to remove Kepone and other pollutants.[67]

With these pieces of evidence against Allied, Cummings felt confident. "And then when the evidence was assembled to the point where we thought we had all that we needed to convict them—not just to indict but [also] to convict them—that's when we asked the grand jury to vote to indict, and they did."[68] The jurors rendered two sweeping indictments on Friday, May 7, 1976. The first charged Allied with 941 counts: 456 violations of the Refuse Act, 484 violations of the Federal Water Pollution Control Act (or Clean Water Act) for dumping Kepone, TAIC, and THEIC without permits, and one count of conspiracy to violate federal pollution laws among Allied and five employees. Each of the five employees, Hundtofte, Frank L. Piguet, James G. Sawyer, Joseph A. Smith, and Gerald P. Williams, was also charged with one count of conspiracy. The second indictment charged Allied, Life Science, Hundtofte, Moore, and the City of Hopewell with 153 counts each of violating the Water Pollution Control Act for discharges of Kepone waste into the city sewer system. Hopewell also was charged with three counts of failing to report Kepone waste in its treatment facility. A third indictment came on August 2, one count of conspiracy among Allied, Life Science, Hundtofte, and Moore relating to discharges of Kepone from LSP. The possible jail times and fines for the defendants if convicted with maximum penalties were staggering. Allied faced a possible fine of $17 million; Life Science, Hundtofte, and Moore $3.8 million each; Hopewell $3.9 million; and each alleged conspirator $10,000 on each count. Each count of discharging without a permit came with maximum jail terms of one year on each count (that could have been 1,093 years). It was up to five years on each conspiracy count. All defendants initially pleaded not guilty.[69]

Over the coming weeks, as the United States celebrated its bicentennial, Merhige presided over not only the lawsuits filed by the former Life Science workers but also those from the seafood industry and now the federal cases. With federal indictments rendered, some of those cited by Cummings began to plea-bargain. The City of Hopewell went first and made history. Among the evidence in Cummings's possession were "handwritten notes by William Havens, superintendent of Hopewell's sewage treatment plant" that showed that Havens accepted for the city discharges of Kepone-contaminated wastewater from Life Science. Further, "He did so knowing that the discharges contained far more Kepone than the 100 parts per billion allowed in the city's amended permit." After the mayor and city council consulted in early June with their lawyers, on June 25 the City of Hopewell officially pleaded nolo contendere (no contest) to ten criminal counts on the second indictment agreed to pay a $10,000 fine for dismissal of the remaining charges. Merhige also put the city on five years of unsupervised probation.[70] It was the first time a city had been convicted of such charges and placed on probation. Afterward, Merhige thought his decision was rather lenient. As he said in 1995: "I might add that the one who got away easy was the City of Hopewell, who were a party to the whole thing. They slipped in right quick and pled quickly and I put them on probation. Everybody thought that was kind of a strange thing to put a city on probation, but I did. I guarantee you, I intended to do something about it if they had violated the probation."[71] The nolo plea has some advantages. For prosecutors it avoids the delay and expense of a trial in exchange for some type of lesser charge, an admission of responsibility, and assistance with other cases. For defendants, conviction on a nolo plea cannot be used as evidence in another legal proceeding on the same evidence. In the Kepone cases, this applied to the civil suits that were underway.[72]

After the plea from Hopewell, the next issue for Merhige was the felony count of conspiracy against Allied for providing false information to the United States. This stemmed from the permits for waste discharges when Allied made Kepone along with TAIC and THEIC. Allied asked the court to dismiss the charge, and on July 6 Merhige agreed on the grounds that a company cannot be in conspiracy with itself. A proper conspiracy, he said, "must accuse an individual or corporation of conspiring with another." Corporations act through their officers or employees and thus were not separate entities. Since the employees were acting as part of their duties, there could be no charge.[73]

In August 1976, as Merhige was busy dismissing Allied's request to subject the Life Science workers to more testing at Duke University, several more pleas came in. On August 10 Virgil Hundtofte pleaded no contest on 79 of the 153 counts against him in the second indictment, and the government dismissed the other 74 counts. He pleaded no contest on the third indictment of conspiracy from August 2. He also agreed to a guilty plea on a lesser misdemeanor charge of conspiring to furnish false information to the federal government for his role with Allied. In exchange, he agreed to testify against Allied and, at least for a time, against his former partner Moore and his former colleagues at Allied. On the same day, Moore and Allied both went the opposite direction from Hundtofte and entered "not guilty" pleas on the third indictment. Moore also entered an innocent plea on behalf of Life Science on August 16.[74] Over the next two days, two more of the five charged with crimes at Allied came forward. Joseph A. Smith, who had been the "technical superintendent" of Allied's Plastics Division, pleaded guilty to a lesser charge of making a false statement to the EPA and the Corps of Engineers. Merhige dismissed the felony charge against him. He agreed to testify for the government. The next day, James G. Sawyer, who had been the technical superintendent of Allied's Agricultural Division in Hopewell pleaded guilty to a lesser charge of "aiding and abetting the illegal discharge of Kepone" in exchange for dismissal of the felony charge.[75]

Then perhaps the largest domino fell. After Allied failed to reach a plea deal with Cummings, on August 19 the company entered its own nolo plea on 940 counts from the first indictment. Cummings vigorously objected, demanding a full trial, and asked Merhige to reject the plea. But Merhige accepted it, noting that it was an admission of guilt. But it also could not be used against the company in civil liabilities pending in other cases. Merhige also noted that Allied still faced the possibility of a hefty fine on the 940 counts and a trial on more than 150 charges related to the company's role in Life Science's actions. There was bound to be enough exposure made public at those trials, he assured Cummings. The decision to plead no contest came against the initial advice of Allied's outside lawyers, Joseph M. Spivey III and Murray Janus. As they planned Allied's defense, they received a call from corporate headquarters directing them to enter the plea. Janus felt they could have persuaded the government to drop half the charges. "But we were told to forget the legal maneuvering. Allied was losing the public relations battle and wanted to put an end to most of the controversy."[76]

The two remaining defendants from Allied, Frank Piguet and Gerald Williams, refused to enter pleas and went to trial in Alexandria on August 31. (Merhige agreed to move the trial from Richmond because the wife of one of the government lawyers was about to give birth). They were acquitted of conspiracy charges on September 2. Merhige stated that the government failed to meet burden of proof. Hundtofte testified at the trial; he did not believe they were engaged in a conspiracy and did not believe Allied was either. Apparently Merhige was also concerned about the blame being shifted to lower-level employees and leaving Allied off the hook.[77] Cummings gave in a little in these cases. He would state in 1995, "We felt badly that two had pled guilty and two were acquitted with a trial. So, the Justice Department authorized us, in an unusual situation I think, to dismiss the counts against the two who had not yet been sentenced but had pled guilty."[78]

There was one more legal move. As the trial loomed for the conspiracy charge against Allied, Life Science, and William Moore for discharging waste, on September 14 Moore changed course and pleaded no contest for himself and on behalf of Life Science to 153 counts of water pollution violations and no contest to the conspiracy charge against Life Science. The government dismissed the conspiracy charge against him.[79] But what remained was Moore's refusal to admit a conspiracy with Allied. Apparently he rejected offers by the government to drop half the charges in exchange for his confession.[80] This would prove helpful to Allied's defense later. After all the plea bargaining, Allied remained the only defendant standing once the federal trial began, and Cummings remained focused on getting a guilty verdict.

The Allied Trial and Verdicts

Setting aside for the moment the 940 counts to which Allied pleaded no contest, the federal government went forward and charged Allied with 153 counts of aiding and abetting Life Science's illegal discharges and one count of conspiracy with Life Science to illegally discharge waste. Was Allied criminally responsible for the actions of Life Science? The key issues Cummings and the government had to prove were that Life Science was an "instrumentality" and agent of Allied, that Allied was an accomplice to Life Science, and that the two companies engaged in a conspiracy to violate water pollution laws. As Merhige promised, the trial exposed many of the actions of Hundtofte, Moore, and others, along with the relationship between Life Science and Allied. And while the "no contest" pleas seemed to have been achieved easily, the case tried in September proved more difficult. Surrounded by "massive notebooks and two portable bookcases containing a dozen binders with research material," Cummings argued that the tolling arrangement between Allied and Life Science constituted a way for Allied to circumvent pollution control laws. Allied was the principal and Life Science the agent. Cummings's arguments represented an effort to "pierce the corporate veil" and show that these two distinct corporations were "so involved and entwined in each other's operations that they [were] really one and the same enterprise." Cummings also needed to prove that Allied at least had "tacit acceptance," if not "explicit condonation," of Life Science's actions.[81]

The trial opened on September 27, 1976, with the prosecution calling Moore to the stand. Moore began by taking the court through the process of how Life Science operated. The company was wholly dependent on Allied to purchase its Kepone, and he noted how the contract required Allied to supply all the chemicals needed to produce the pesticide and to cover Life Science's expenses for equipment. Only two pieces of equipment came directly from Allied. At one point Merhige asked Moore: "Did Allied run Life Science Products or did you run it?" Moore replied: "The relationship was very close." But he added, "[No one at Allied] told us to wear our ties sideways."[82] Moore frequently contradicted himself. He first testified that he had continuously informed Allied of Life Science's failure to meet pollution requirements. But defense produced a letter from him to Allied dated March 1975 in which he claimed, "[At] the present time, Life Science appears to be complying with regulatory agency requirements." In that same letter he noted that the company was looking ahead to the next contract year with Allied and planned to expand production capacity to ten thousand pounds per day and requested more money from Allied to cover the costs of the expansion.[83] One of Allied's lawyers, Murray Janus, later recalled that it was easy to get Moore to contradict himself, since he had made statements at the Senate and House hearings and been deposed by lawyers. "If Moore mentioned safety goggles," Janus explained, "I could flip to six previous statements on goggles. As often as not he had said six different things, all under oath. Cross-examination was a piece of cake." It helped Janus that Moore had refused the government's offer of a plea bargain on conspiracy because, as Moore stated, "I felt on my part there certainly was no conspiracy."[84]

Hundtofte's testimony filled most of day two. He confirmed Moore's recollections that they called the meeting in July 1975 with Ike Swisher of Allied after writing him the month before that Life Science was not going to meet the new pollution requirements. Allied did not make a move to shut down Life Science once its executives learned about the pollution. Indeed, both Hundtofte and Moore indicated that, if anything, Allied wanted Life Science to produce more Kepone.

Hundtofte said that Allied would pay for any new equipment, which Ike Swisher confirmed, but only after Allied approved the plan. Hundtofte gave contradictory testimony on whether Allied continued to supply materials to make Kepone after this meeting—the key date for forming the conspiracy according to prosecutors. Allied's lawyers continued to argue that the company could not have been conspiring to pollute if the company had been told that Life Science complied and then, once Life Science admitted it could not meet the requirements, offered to pay for new equipment. Hundtofte admitted that Life Science had failed to limit its discharges, and he read from company logs that indicated that at times tanks of the pesticide were dumped into the sewer system without treatment.[85]

Other testimony followed regarding the haphazard operations at Life Science, including those of Del White, Dale Gilbert, and Bill Moyer, who took the stand still suffering from Kepone shakes. White pointed to how the goal as of August 1974 was to double production from three thousand pounds per day to six thousand. Gilbert said at times the plant had produced as much as twelve thousand pounds on one or two days, "but that meant more effluent. We weren't able to handle it." Untreated Kepone waste was often dumped or flushed into the sewers, hundreds of pounds of it.[86] An internal memo written by a technical service representative from Hooker Chemical, which suppled the HCP used to make Kepone, recounted two visits to Life Science and stated that HCP was in the air around the plant and on the ground in large amounts, along with Kepone powder everywhere. Hooker sent the memo to Allied, making this at least three warnings Allied received about conditions and having been received by Hundtofte and Moore who said as much at the meeting with Allied in July 1975. Next on the stand, H. D. Howard recalled his meetings with SWCB officials, who did not close down the plant even after being "disturbed" by the amount of Kepone being discharged. William Havens, the superintendent of Hopewell's sewage treatment plant, next recalled figuring out that it was Life Science causing the digester problems by following the smell from a pump station back to the plant on Randolph Road.

Hopewell's city manager at the time of the events, Arthur Lane Jr., testified that he never informed federal officials that the city's sewer system discharged Kepone. Merhige pushed the analysis of the city's role in the tragedy. He asked Hundtofte: "I gather more often than not you were telling them you were violating their permit allowing the company to use the city sewage system?" "Yes, sir," replied Hundtofte. Allied's lawyers used statements like these to shift blame to state and local authorities who allowed the waste to enter waters in violation of the permit. Merhige grew frustrated at the lack of response from the state and city. As John Reeves was testifying about his various meetings and memos on the issue, Merhige interrupted him: "It sounds like a bunch of politics, everybody being nice to everybody else."[87]

On late Wednesday afternoon on September 29, Merhige granted the defense motion to dismiss the 144 counts in the first indictment of aiding and abetting Life Science, before the company even presented a defense. Prosecutors had not provided enough evidence "to establish criminality on the part of the defendant." Merhige likened Allied's responsibility to that of a prescription drug manufacturer who has a legal duty to warn physicians of the potential dangers of a drug. After that initial warning is given, further responsibility then rests with the physician to warn and inform others. Because Allied had notified Life Science of the nature of Kepone, Allied had no further duty to report to the Environmental Protection Agency or any other regulatory body. Merhige concluded that the ultimate fault for the illegal discharges lay with Life Science, thus Allied could not be found guilty of criminal actions performed by Life Science. Yet Merhige remained concerned with Allied's actions after the July 7 meeting with Life Science, and he allowed testimony to continue the next day on the nine remaining charges covering the period from that meeting until Life Science closed on July 23. There was also the outstanding conspiracy charge against Allied from the third indictment.[88]

Overnight, Janus and the other Allied attorneys debated about whether to call any witnesses at all and take their chances on acquittal on the remaining charges, given the apparent weakness in the prosecution's case. Allied's president apparently left the decision to Janus, and he decided to call only two witnesses, Moore and Patrick L. Henry, the Allied lawyer who negotiated the contract with Life Science and who was present at the July 7 meeting. "I decided to hedge our bets by offering just a very small part of defense case," he recalled. Just "enough to tip the scales and push Merhige toward an acquittal."[89] Moore reiterated that there was no agreement between Life Science and Allied at the meeting to discharge at a higher rate than permitted by Hopewell. Henry added to Moore's testimony by noting that after the meeting, he was convinced that Life Science would be able to meet the pollution goals with Moore's new treatment plan. The City of Hopewell had also required Life Science to install three thirty-thousand-gallon tanks to hold the Kepone waste for testing prior to dumping. Henry had drafted a letter indicating Allied's willingness to pay Life Science $151,000 for the equipment. But before the letter could be sent, Life Science shut down. Prosecutors urged Merhige to consider that Allied had a large amount of prior knowledge of discharge problems, including testing effluent for Life Science that showed higher than permissible levels of Kepone. Allied knew of the problems at the July 1975 meeting and could have stopped them simply by refusing the accept delivery of Kepone. Janus admitted that Allied could have done more to stop the pollution but argued that negligence is not enough to establish guilt of a criminal conspiracy.[90]

After closing statements, Merhige secluded himself in chambers for almost an hour. He returned "to the walnut-paneled courtroom" at 3:55 p.m. He reminded the lawyers that there is a presumption of innocence that is "not easily pierced" in criminal cases. He leaned on his elbows to review the evidence and then pronounced that it left him with a reasonable doubt and that it was not sufficient to show that Allied had knowingly aided and abetted Life Science's illegal discharges. "I have no choice but to acquit," he said, acquitting Allied Chemical on all remaining counts. It was a significant victory for Allied. But, Merhige reminded, there remained the 940 counts to which Allied had pleaded no contest, and a firm ever convicted of pollution charges in his court would know it. With that, the court adjourned, and Allied's lawyers and executives hosted an all-night celebration, the bill for which was "rumored to be the largest in the history of Richmond." No doubt they could afford it. *Fortune* magazine wryly noted that Kepone litigation "was the biggest windfall to hit Virginia's legal industry since personal-injury suits were invented."[91]

While Merhige acquitted Allied on the conspiracy charges, he clearly had been moved by the poisoned environment and the plight of the workers, including Dale Gilbert and Del White who shook from tremors as they testified. Merhige also continued to hear this federal testimony as he reviewed evidence in the civil cases filed by the Life Science workers. So perhaps Allied and its lawyers should have been ready for Merhige's October 5, 1976, bombshell ruling on the 940 counts, which set a precedent for the fine and its imaginative solution. Before announcing the fine against Allied, Merhige finalized other remaining matters. Sitting at the bench, he first fined Life Science over $3.8 million on 154 charges, a symbolic move since the company was insolvent. He also fined the two owners of Life Science, William P. Moore Jr. and Virgil A. Hundtofte, $25,000 each and placed them on five years' probation. And, at the government's request, he dismissed the remaining misdemeanor conspiracy charges against Hundtofte, Sawyer, and Smith. When it came time to sentence Allied, Cummings argued for the maximum and introduced witnesses to bolster his case. They included David Bailey from the SWCB, who stated that he had tested bluefish frozen in 1973 for Kepone, and they contained levels higher than the action level. The state had no reason to test for Kepone before, since it was an unknown compound. Cummings said that Allied had shown no remorse for its actions, and, had not Life Science come to public knowledge, Allied's actions and the presence of Kepone in the water would have gone undetected. Janus appealed to Merhige to consider Allied as a "good corporate citizen" and impose only minimal fines for TAIC and TEIC and impose maximum fines only for Kepone.

By this time, investigations and news reports emerged about how widespread Kepone had become. Trace amounts of Kepone were found in mothers' milk from nine women in seven cities in the southeastern United States, likely from Mirex, used heavily against fire ants, which breaks down into Kepone.[92] News from Baltimore showed Kepone on the grounds of the Allied plant there. Fish from the lower James Bay sold in markets in Virginia, as well as Baltimore, Philadelphia, and New York, contained Kepone, some above the action level of .1 ppm, prompting some fear among consumers. Then, in July 1976, tests showed some bluefish in the Chesapeake Bay contained Kepone above the action level. While the fishing ban remained largely in effect for the James, Governor Godwin did not close Virginia's section of the lower bay.[93]

With all these events swirling in the background, Merhige deliberated over the sentencing and then returned to the bench to announce his ruling. Reflecting his own and the larger public concern for the environment, Merhige began by stating, "[Pollution] is a crime against every citizen. The environment belongs to every single person, every single citizen, from the lowest to the highest." He did agree with Janus that Allied had been a "good corporate citizen." Yet he disagreed with "the defendants' position that all of this was so innocently done, or inadvertently done." "I think it was done," he said, "because of what it considered to be business necessities, and money took the forefront." He said Allied "knew it was polluting the waters." He commented that after the acquittal of Allied on the earlier charges, "I came away from the case ... with the feeling that the only ones who really cared were some young man in the State Water Control Board, his immediate superior who kept after him to bug Life Science, and the State Health Department." His general view, Merhige said, was to "temper justice with mercy." Then he asked, "How do we do it?" The answer followed: he announced that Allied would be fined the maximum amounts of $2,500 for each count 1 to 456, and $25,000 for each count 457 to 940—$13,215,000 in total. At the time, it was the largest fine ever imposed in a federal pollution case. But he also suggested that it would be better if the money, rather than going to the federal treasury, "could be used to benefit those who directly were hurt." He would not compel Allied to do so, nor would he reduce the total amount of money, but he would consider any actions "taken voluntarily" by Allied.[94]

Janus criticized the decision. "It's the only time I've ever been upset with Judge Merhige," he said. "He's a conscientious judge who normally agonizes over sentencing. Thirteen million dollars was the largest fine ever imposed at that time, and I think Judge Merhige paid too much attention to his place in history." "On the other hand," Janus admitted, "Judge Merhige did give us a fair shake on the conspiracy charge. His potential place in history didn't color his objectivity, and he didn't hesitate one bit to enter an acquittal when the evidence was insufficient." "Still," Janus reflected, "I took the sentence personally because we worked so hard to show that Allied was a good corporate citizen. Judge Merhige just wouldn't listen." Merhige handled the criticism with humor. When later introducing Janus at a bar function, Merhige joked: "I made Murray famous. I gave *him* a place in history by hitting him with the largest criminal fine ever recorded. Although I'm not in the habit of explaining my decisions, I think you are entitled to know why I imposed that fine. I fined your client $13.24 million because the law would not permit me to fine them $13.24 million and one dollar." In a later interview, he explained that pollution was an indiscriminate crime that kept circulating, "hurting more and more people." Appropriate punishment should follow: "I believe in stiff penalties for such indiscriminate crimes."[95]

An Imaginative Solution: The Virginia Environmental Endowment

In deciding how to implement Merhige's desire to keep some of the money in Virginia and reduce its fine, Allied looked for ideas. They may have developed what became the Virginia Environmental Endowment out of separate negotiations with Virginia officials on their pending lawsuits. Otis Brown recalled that he suggested an idea to Allied vice chairman Sandy Trowbridge when meeting with him. "And so I said, 'I want to set up a foundation, and I want you to finance it. So, we're going to help this seafood industry get back in business.' And he agreed to it. And I've forgotten all the dollars now, but we spent all our morning working on that. And [Trowbridge] said to me, 'We got court cases this afternoon with Judge Merhige. Do you mind if I share with Judge Merhige what we're going to do with you?' And I said, 'No problem at all.'"[96] Later, when the Virginia lawsuits were settled, some of that money did eventually go to help the seafood industry, but it's not clear if there was a separate foundation as Brown recalled.

What is certain is that on January 28, 1977, in middle of a brutal winter that led to fuel shortages, school closings, and rising unemployment, Allied presented Merhige with a proposal. In lieu of the full $13.25 million fine, Allied offered instead to spend $8 million to create an environmental endowment and have the actual fine be only $1.44 million. Allied came to the amount for the endowment "after discussion with state and Hopewell officials about possible Kepone-related projects." The endowment would "use funds to improve and enhance the environment in and about the commonwealth of Virginia."[97] After calling Allied's offer "very generous," Merhige considered it for a few days. Then, on February 1, Merhige agreed to the proposal but kept Allied's fine at $5 million, to maintain the total of Allied's financial commitment closer to $13.25 million. The $8 million became the basis for the Virginia Environmental Endowment (VEE), a silver lining from the dark cloud of Kepone. It became the first "grant-making foundation in the nation to devote its funding exclusively to environmental issues."[98] In the 1980s, the VEE received money from other legal settlements as well. Since its founding, the VEE has funded over fourteen hundred grants in a range of environmental areas, making a significant impact on Virginia's environment. At the time, Merhige stated that the company had become "good guys" in his book for its response. In a statement announcing the deal, Allied head Connor said that he told Governor Godwin that Allied "would do the right thing in responding to the environmental threat posed by Kepone." Of course, the endowment would not have happened without Merhige's prompting.[99] Cummings remembered that the Allied representatives came to court "with a check for eight million dollars payable to The Virginia Environmental Endowment, and I guess they had a blank check in their back pocket because they didn't know what the fine was going to be. Judge Merhige then reduced the fine by about that amount of money. So, they wrote a check out for five million and some dollars."[100]

Merhige maintained control over selecting the board for the new VEE. His first pick was Cummings. Other initial board members were Ross Bullard, a U.S. Coast Guard rear admiral in charge of the mid-Atlantic region; Cathleen Douglas, lawyer and wife of Supreme Court Justice William O. Douglas; Frances A. Lewis and Sydney Lewis, founders of Best Products and noted philanthropists in the Richmond area; Henry W. MacKenzie Jr., a circuit court judge from Portsmouth, Virginia, knowledgeable on water law; and George Yowell, president of Metropolitan National Bank in Richmond. The first three areas of priority for grants were toxic substances, environmental law, and ecological research.[101] Anthony Troy, who became state attorney general in January 1977 when Andrew Miller resigned to run for governor, urged Merhige to give priority to Kepone. "A substantial portion of the funds of the endowment should, in the final discretion of the board of directors, be directed to implementing those programs which research shows would be most promising."[102] Troy's office also asked for a share of the $8 million. But Cummings and the board were adamant in not paying the state government money for Kepone. As Cummings said: "Our first order of business was to tell the attorney general of Virginia that he couldn't have the money, as I recall." The board followed Judge Merhige's directions. The money, Cummings said, "was designed to help the citizens of Virginia, in a way that government could not, through its own restrictions and limitations." "Then," he recalled, "the next order of business was to hire Jerry McCarthy."[103] McCarthy was still head of the Council on the Environment, which, under Governor Holton, assumed a leading role in promoting Virginia's environmental policy. As McCarthy recalled years later, the VEE board members "were very smart people with wide knowledge inventories," but environment issues were not their forte. "And so they then said, 'Well, what we need is someone who understands the issues.' And that wound up in this case being me."[104] McCarthy would go on to lead the VEE for over three decades before retiring in 2013.

Fines imposed as part of criminal sentencing are the main tool courts use against a company. Companies cannot go to jail. What effect did the fine have on Allied? Certainly, Kepone brought bad publicity to Allied, and the size of the fine warranted attention. Allied did get a tax deduction on the $8 million put toward the environmental endowment, though, reducing the net cost of the Kepone outcome. With legal expenses and voluntary contributions to the cleanup and study of Kepone, estimates of the total cost to Allied are over $30 million.[105] But $13 million (or $30 million) is a small price for a company with some $3 billion in annual sales in 1977. The Allied annual report for 1977 indicated that the company's net income rose 15 percent and its share price increased, despite the "extraordinary charges relating to the Kepone problem at Hopewell, Virginia." Its corporate image did not suffer, since its operations were diverse, and its customers were other businesses unmoved by Kepone.[106] One of Allied's lawyers, Robert Sand, noted that internally Kepone forced changes to the company's procedures. One lesson, Sand said, was that "if you're going to toll process a chemical which is toxic, either give it to a big responsible company, or if you use a small company, watch them like a hawk."[107] Allied tightened its procedures on contracting, upgraded environmental affairs to the vice president level, instituted a Toxic Risk Assessment Committee, and provided incentives to motivate personnel and executives to achieve environmental goals. The change warranted praise from the EPA and also the *New York Times*, which editorialized: "Allied Chemical deserves recognition for an impressive corporate turnabout."[108]

Settling with the Seafood Industry

As Merhige was busy sorting out the workers' legal cases and the federal case against Allied, he also heard civil cases from the seafood industry. The fishing and harvesting ban led commercial fish and shellfish operators, oyster tongers, clammers, crabbers, and marina owners to file civil lawsuits. There were two major cases, *Adams v. Life Science* and what became a complicated class action lawsuit, *Pruitt v. Allied*. On January 20, 1976—just days before the federal grand jury would begin investigating Allied—120 commercial fishermen, oyster bed owners, tongers, and crabbers filed the *Adams* case, claiming $2.7 million in damages from Life Science, Allied, Hopewell, Huntofte, Moore, and Norwood Wilson, owner of the former pebble plant land where Life Science workers dumped Kepone waste. The complaint stated that "the James River was converted by Allied and Life Science into their private sewage disposal system." It also stated, "Kepone has destroyed the market for any seafood from the James River and its tributaries causing economic losses and damages to the plaintiffs."[109] Others joined the suit so that the total claimants reached over 250 and damages increased to over $16 million.[110] Eventually almost all of the parties settled in June 1980 for an amount stated to be about $500,000. One group of fifty plaintiffs said they were sharing $180,000, so the total amount seems likely.[111]

The *Pruitt* case, often cited since in tort and maritime law and taught in law school, was a class action lawsuit filed a year after *Adams*, on January 24, 1977. A group of twenty-six Virginians and Marylanders filed the case on behalf of thousands of others from both states who made their living from the water. They sought status as one class, people whose "livelihood or income is derived from, or dependent upon, the catching, taking, buying, selling, processing, packing, packaging, or distributing of seafood from the Chesapeake Bay, the James River, their tributaries, and adjacent water areas."[112] Merhige first had to determine if all the plaintiffs warranted certification as one class. It took until January 3, 1980, to sort this out, as the case proved more complex than first realized. Rather than one class, Merhige instead divided them into three. The first, "commercial fishermen" included fishers, shellfishers, and oysterbed lessors. Merhige went further here, though. The plaintiffs claimed to represent both Maryland and Virginia watermen. Merhige, citing the history of the "oyster wars" and ongoing antagonism between these groups, refused to allow this. He also noted that differing state laws made it problematic when it came to maritime claims. The "Court deems it prudent to exclude all plaintiffs who reside in Maryland from this and all other subclasses."[113] The second class, "distributors," included seafood enterprises not engaged in fishing operations, such as wholesalers, retailers, processors, distributors, and restaurants. Finally, the third class, "surrogates," included businesses connected to the sportfishing industry, such as boat and marina owners and tackle and bait shop owners who could press their own claims and act as surrogates for recreational fishers.[114]

Another challenging aspect of the case was that the groups were not claiming direct property damage but rather sought restitution for lost profits because of the decline in demand for tainted seafood. Since the fish and oysters did not die, no property had been damaged, and only market demand changed, Allied sought to dismiss the claims.[115] Merhige, though, looked to the damage to the river and the bay as a critical issue not to ignore. He favored the claims of the commercial and the sportfishing groups. "The fact that no one individual claims property rights to the Bay's wildlife could arguably preclude liability. The Court doubts, however, whether such a result would be just. Nor would a denial of liability serve social utility: many citizens, both directly and indirectly, derive benefit from the Bay and its marine life. Destruction of the Bay's wildlife should not be a costless activity."[116] Indeed, Merhige argued, "The Court perceives no valid distinction between recognition of commercial damages suffered by those who fish for profit and personal harm suffered by those who fish for sport." In keeping with traditional understandings of maritime case law, Merhige refused to accept the claims of the second group, the distributors, seeing them as being too far removed from the water's edge to receive damages. He again sought to balance the threat to the marine environment with the interests of all parties, to "tailor justice to the facts" of the case, and he urged the parties to settle. They finally did so in June 1982, five years after the case began.[117]

Settling with Virginia

The fourth major legal challenge came from the State of Virginia, which sought restitution for costs associated with Kepone. On Friday, February 13, 1976, the SWCB authorized the attorney general to begin civil damage proceedings against Life Science as well as Allied and Norwood Wilson Jr., who owned the pebble plant property where Life Science dumped Kepone waste. On March 22, 1976, as the "Kepone bills" met approval in Virginia, the state's attorney general Andrew Miller, acting on behalf of the SWCB, filed civil suit for a little over $3.5 million against Allied, Life Science, Moore, and Huntdtofte. They sought "maximum penalties" for violations of state water pollution laws on five counts between February 21, 1974, and July 23, 1975. Each count charged the owners, including Allied, with failing to make an application for the discharges or provide facilities approved by the SWCB for treatment and control. Over half of the suit, $2,040,000, addressed dumping Kepone at Norwood Wilson's pebble plant for 204 days: a fine of $10,000 per day. Another $950,000 was linked to violations of state waste discharge procedures from July 1, 1974, when the procedures came into effect, until October 3, 1974, when the SWCB first became aware of Life Science's existence. The third count addressed waste that leaked through a crack in pipe at Life Science and sought $510,000 for fifty-one daily violations. A fourth count sought $40,000 for Life Science's dumping of waste at the Hopewell landfill, and the final count sought $10,000 for Life Science waste overflows from a quench tank.[118]

Meanwhile, out of the public eye, Virginia's Kepone task force continued discussion with Allied, the attorney general's office, and Governor Godwin. State officials were trying to convince the company to settle and pay. Jim Ryan was the point person for the attorney general's office and remembers meeting for more than a year to address issues raised by their settlement demand. "We basically said, 'You were the real party in interest here. You provided the raw materials. You got the finished product back. You sold it. All they were doing was converting it for you, and you can't outsource your problems like this. Your chairman told the governor that Allied wanted to be a good corporate citizen in Virginia, and you're going to make this right.'"[119] The state's initial price tag was somewhere between $21 and $25 million. This included addressing the remaining Kepone and the 1.3 million gallons of Kepone sludge sitting in an asphalt-lined lagoon at the sewage treatment plant in Hopewell, studies of Kepone in the river and in the bodies of the workers, and studies on how to lessen the economic impact on the seafood industry, especially watermen put out of work. The state also requested money for watermen and to build an incinerator to burn Kepone waste.[120] During regular talks with the company, Godwin too was insistent on getting Allied's help. He had from the beginning talked directly with their leadership to "make things right," as he often said. He told Allied vice chairman Sandy Trowbridge in August 1977 that "the Commonwealth had not received the first cent from Allied, and that, in Virginia, Kepone meant Allied."[121] Otis Brown was involved too. He resigned from government in May 1977 but stayed on without pay at Godwin's request to help with the Kepone crisis. He remembers Allied being "excellent to work with" and "very positive." "At no time did Allied ever back off and say, 'Listen, we're not going to do that.'"[122]

Ultimately, on October 13, 1977, Allied settled and paid Virginia and the City of Hopewell $5.25 million for additional cleanup. This was in addition to the money Allied had already paid out for cleanup at Life Science, among other costs. In agreeing to the settlement, the state dismissed its lawsuit against Allied. The agreement left the state able to seek further damages after three years for removal or containment of Kepone but limited Allied's liability in paying for monitoring and research costs associated with Kepone. Virginia abandoned the plan for an incinerator to burn Kepone waste as Allied balked at paying for that (its cost estimated at between $7 and $10 million) and abandoned additional money for watermen. The state and Hopewell also received a guarantee that Allied would not make claims against them. The moratorium allowed the state time to examine various options for removing Kepone from Hopewell and the sediments in the James. Hopewell received $650,000 to pay for damage to the city's treatment plant. Godwin praised the agreement as "evidence of earnest efforts of the commonwealth in cooperation with Allied to move this unfortunate Kepone story toward a close."[123] Allied also funded further research on Kepone at the Medical College of Virginia and paid the costs of cleaning up the Life Science property.[124]

The legal proceedings over water and those entailing the workers formed another important aspect of the residues of Kepone. These proceedings are part of the "residual governance" of toxic events like Kepone, as are the cleanup efforts, monitoring of waste, and medical studies.[125] The court decisions handed down by Merhige drew on earlier precedents of environmental contamination and set new ones, further encoding the past into present legal and regulatory systems. Moreover, the Virginia Environmental Endowment is both a living embodiment of Kepone's toxic legacy and an organization that funds projects addressing other residues of environmental damage. Meanwhile, as these proceedings moved forward, government officials busily drafted new laws and regulations to address the Kepone crisis and toxic substances generally, further evidence of residual governance. As expected, these new proposed laws and regulations led to fierce political battles, especially over the closure of the James River. These battles showcased the scientific uncertainty surrounding the effects of Kepone in humans and the growing backlash in the 1970s over government regulation.

CHAPTER 6

Kepone and the Environmental Management State

The Virginia Department of Health (VDH) website contains a "Fish Consumption Advisory" that provides information to help anglers make "educated choices" about eating the fish they catch. The advisory is not a prohibition but "a warning about the contaminants present in a fish species." There are recommendations for specific fish as well. One can search by body of water or specific contaminant. There, under "James River," is Kepone. It is found in all species, "from the I-95 James River Bridge to the Hampton Roads Bridge Tunnel." The VDH recommends eating no more than one meal per day of James River fish.[1] Like the Kepone extant in marine life, the website and list are both residues of Kepone and the twentieth-century dependency on chemicals and the regulatory state that emerged along with them. Behind the website and warnings lies a history of how those chemicals entered the environment and how Virginians used and understood those chemicals and their effects, debated, and responded to them.

Kepone's most immediate legislative impact in Virginia was the state's Toxic Substances Information Act. At the federal level, the events in Hopewell added support for the Toxic Substances Control Act (TSCA). The Kepone incident also played a role in the 1977 amendments to the Clean Water Act. While each piece of legislation produced some opposition, more vigorous debates emerged over the fishing ban in the James River, centered on the interrelated questions of whether Kepone was a danger to human health, whether there was a safe level of consumption, and how policy makers determined those levels. The laws and the ban emerged during a growing national concern over cancer in the 1970s. Scientific data on Kepone showed its hazards as a potential cancer-causing agent, and the visible effects on workers' bodies made it clear that the pesticide could be harmful. As pesticides and other chemicals became more widespread, advances in technology allowed measurement of these substances in ever smaller quantities. With Kepone, regulators initially did not even account for it. Then as they began to develop testing mechanisms to find it in bodies and the environment, public health officials and others could begin to assess its risk and bring Kepone into the growing register of monitored substances. Rather than issue an outright ban on the consumption of fish and shellfish with Kepone in them, the FDA and EPA relied on the idea of a threshold effect for chemicals, setting "action levels" that determined the amount of Kepone consumers could safely eat in certain species.

Regulators faced a growing conservative challenge that sought to limit public transparency in chemical production and exploited the uncertainty over the toxicity of various chemicals to limit regulations on toxic substances. This dovetailed with individuals along the river who disputed the dangers and ignored the regulations, fishing and harvesting crabs in defiance of the ban. Watermen and their trade associations, their lawyers, scientists, and political leaders worked in concert to question animal testing and what those tests showed for the potential long-term dangers of Kepone on humans, especially cancer. They argued for the general safety and necessity of pesticides and other chemicals. They sowed doubt and uncertainty, using arguments and strategies made famous in opposing regulations on tobacco and asbestos, and those related to acid rain, the ozone hole, and climate change.[2] In Virginia, as political pressure grew, policy makers continued to evaluate the ban on the James River. Initially they valued prudence in the face of ignorance about Kepone and responded using data from studies on Kepone and adherence to the action level, not political expediency. It appears that, even before Kepone levels began to fall, some in Virginia's Health Department and the governor joined in pushing for a change in action levels. Still, the ban did not end until 1988, when Kepone levels in marine life dropped below the action levels set in 1977. Support for the measures associated with Kepone reached across party lines. The legislation had weaknesses, earning some deserved criticism. But amid Kepone's uncertainty, regulators and political leaders leaned initially toward the precautionary principle when it came to protecting human health.

These events take us inside the construction of the environmental management state. In the post-World War II era, environmental concerns became "a mass phenomenon" with a deep impact on politics.[3] The Kepone affair raised questions about government and corporate responsibility in matters of public health, environmental protection, and worker and consumer safety.[4] In Virginia, poisoned workers and a contaminated river proved powerful examples of the need for an expanded public role in protecting human health and ecosystems regarding toxic substances. Yet the events associated with Kepone also illustrate the way "risk as an organizing principle" developed in regulating hazards such as toxic substances. In this case, the environmental management state emerged during a time of heightened concern with a substance new to the public and entailed attempts to assess Kepone's risks to marine life and, even more so, human health. Kepone and other hazards reflected the "contradictory nature of modern science: as both a source of dangers and a chief means by which we can detect and understand these dangers," in the words of Stephen Bocking *Nature's Experts: Science, Politics, and the Environment*. Evident as well, Bocking notes, was the demand to "set priorities by comparing the costs and the benefits of regulating risks."[5] Moreover, the controversy surrounding the ban on fishing reveals the ways that uncertainty and complexity, professional connections, economic interests, and values and assumptions can affect evaluations of risk.[6]

The Kepone Bills: Virginia's Response to Toxic Substances

Kepone spurred a quick response from Governor Godwin's administration and state legislators to craft a series of bills to prevent "another Kepone" from happening. The Kepone task force established to address the effects and cleanup of Kepone now added creating legislation to its workload. At a meeting in January 1976, the task force noted that Otis Brown from Governor Godwin's office had contacted Jerry McCarthy of the Council on the Environment to determine how the Kepone incident happened and suggest "administrative actions to prevent this from happening again." McCarthy surveyed state agencies in the task force and found that they lacked knowledge of products being manufactured, a registration system to make the state aware of a manufacturers and users of toxic substances, and personnel for an adequate inspection program. McCarthy also found that general practitioners lacked training in industrial medicine. There was also no reporting system for occupational illnesses, except for workers' compensation. Another issue was that the federal government had yet to approve Virginia's state occupational health laws, which meant state inspectors had no enforcement authority, having to wait for federal inspectors.[7]

In the Virginia attorney general's office, drafting and getting the bills through the legislature fell to Jim Ryan. "I was the main guy in the attorney general's office, I guess, that was the point man handling all of this on a day-to-day basis." Ryan says that he met with Deputy Attorney General Jim Kulp and Dave Evans: "[We] sat down and identified all of the problems we had in state law that we needed to change."[8] There were ten bills, with nine finding their way into law. The centerpiece was the Toxic Substances Information Act. As Ryan recalls, the "basic purpose was to have industry that used or produced toxic substances disclose to the state that fact and what was known about those substances. Then, hopefully, we would not be ambushed again by such a problem." He drafted the bill along with others and reviewed them with the attorney general and the state's Kepone task force and then presented them to the governor, Health Secretary Otis Brown, and Environment Secretary Maurice Rowe. Jerry McCarthy of the Council on the Environment CEA was likely involved as well. "At a subsequent meeting that included these folks, the governor charged Senator Herb Bateman to patron the bills and get them passed."[9]

Indeed, it was Bateman, Republican of Newport News, chair of the Committee on Agriculture, Conservation, and Natural Resources, who shepherded the bills through. Bateman was born in North Carolina in 1928 and grew up in Newport News, graduating from Newport News High School. After teaching high school, he served in the U.S. Air Force during the Korean War. He then earned a law degree from Georgetown University and practiced law in Newport News. He won a seat in the Virginia senate in 1967 as a conservative Democrat, yet like others in the party (such as Godwin), as the Democrats moved in a more liberal direction, he changed to the Republican Party in 1976. He served in the state senate until 1983, when he began his career in the U.S. House of Representatives. He served until his death in 2000 from effects of lung and prostate cancer. He advocated strongly for his district in the Virginia Peninsula, especially working to ensure defense contracts, which included Newport News Shipbuilding.

In the House of Delegates, floor management of the Kepone package fell to future governor Gerald Baliles, a Democrat who was then a newly elected delegate from Richmond. Baliles had an early interest in the environment. He grew up on a farm, developing a childhood love of the outdoors. "I suppose the real awakening occurred with the reading of Rachel Carson's book, *Silent Spring*, in the sixties, when I was in college."[10] Baliles had worked as deputy attorney general under Andrew Miller, focusing on environmental issues, and served on environmental committees in the legislature. While he was with the attorney general's office, he also served as a go-between for Republican Governor Holton and his Democratic attorney general, Andrew Miller. Baliles also got to know Jerry McCarthy, who was serving as head of the Council on the Environment. Later, as governor, among other achievements he pursued several measures to address pollution in the Chesapeake Bay, and he continued advocating for the bay in his post-gubernatorial career.

In putting forward the legislation that included the Toxic Substances Information Act, Baliles noted, "[The Kepone incident] made us all aware of our responsibilities in the area of toxic substances and how little we know about them." He recalled that those working on the legislation wanted to prevent future Kepone situations: "Governor Godwin asked me as a freshman legislator as I recall ... to handle a very controversial bill that would have required all companies that handle toxic substances to develop an inventory list of types and amounts, storage, transportation, various other things."[11] Godwin had endorsed the Democrat Baliles in his 1975 campaign for the House of Delegates. Working closely with Baliles was Democrat James Samuel (Sam) Glasscock of Suffolk. Baliles remembered that there was "a significant amount of opposition to get it out of committee." When it first reached the floor, Baliles recalled, "[I] recognized the bill was dead on that first vote and wanted to protect it. I don't remember whether it was the next day or the next week, whether it was over a weekend, but several of us worked our friends in the legislature fairly hard, and the governor too, and the next vote was successful."[12]

In March 1976, as demolition crews razed the Life Science operations building and the State Water Control Board was filing suit against Allied, the Virginia General Assembly approved legislation to address toxic substances. All but one of the Kepone bills became law in March 1976—reflecting a bipartisan consensus on the need for them and the success of Baliles, Bateman, Godwin, McCarthy, and Ryan generating support. The House of Delegates killed one bill that would have allowed the SWCB to impose penalties for violations of its regulations; the board did have the power to impose penalties for state law. Opponents argued that the bill would give the SWCB too much power over matters not related to toxic substances. The other nine bills, including the Toxic Substances Information Act, were approved. One gave the courts options to impose criminal penalties or civil fines up to $10,000 for violations of the state air pollution and Board of Health regulations. On these bills, Ryan had Allied square in mind. "When we had earlier filed our State court civil penalties action against Allied Chemical, we were limited in the amounts of penalties we could seek. We wanted to increase those amounts to provide for more deterrence."[13] Baliles also sponsored a bill that would bring Virginia's occupational safety and health laws up to federal standards under OSHA. This would finally allow Virginia to supervise and inspect factories, something lawmakers hoped would deter another situation that had occurred at Life Science. Another bill, responding to the issues related to Life Science workers, required physicians to report occupational diseases resulting from toxic substances to the state health department. Circuit court judges were also allowed to issue search warrants in cases when inspectors were refused entry to plants producing toxic substances. One bill extended from one to three years the statute of limitations for prosecuting violations of regulations covering dumping, discharge, and emission of toxic substances. Another permitted the SWCB to issue cease-and-desist orders without going through courts in cases when there is an immediate safety threat. Still another bill allowed the State Health Department to regulate disposal of toxic substances in landfills. Finally, new legislation required sewage system owners to conduct a survey of industrial waste and report results by July 1, 1977.[14] Each one of these acts addressed some aspect of how government had failed to stop the Kepone contamination from happening and to deter it in the future.

What did Virginia's Toxic Substances Information Act (TSIA) of 1976 do? The premise was that had Virginia regulators known Life Science was producing Kepone, they would have acted sooner to prevent the poisoning of workers and contamination of the river. As Baliles and other supporters noted, the law was informational, not regulatory like the federal Toxic Substances Control Act (TSCA) that will be covered below. Since "few people know what is manufactured" and by whom, the law provided that information.[15] As Jim Ryan explained in 1977, the purpose of the act was to "acquire all information available so that the Commonwealth's regulatory agencies [could] make well-informed decisions on toxic substances."[16] It designated the VDH as the agency responsible for toxic substances information and charged the agency with creating a list "of those substances which the Board has determined to be toxic." Businesses that manufactured or emitted listed substances had a duty to report that to the VDH. The report had to include the names and properties of the substances, how each was emitted into the environment or workplace, and—in a direct reference to the Kepone incident— note whether there was reason to believe that such emissions might harm both waste treatment facilities and employee health. The report needed to include information on what protections existed at the facility to protect worker and public health. The act also required employers to direct workers to physicians for diagnosis of injury or illness suspected as coming from a toxic substance. The law created the Toxic Substances Advisory Council, charged with reviewing and evaluating policies and programs of Virginia with respect to toxic substances, making recommendations to the Health Department, and furnishing technical advice. The governor appointed five members from fields of agriculture, medicine, labor, industry, and local government. The remaining members were from various relevant state regulatory agencies. The new law also required that no trade secrets could be revealed by the VDH or the Advisory Council.[17]

Given the circumstances surrounding the Hopewell Kepone contamination, it is not clear that these measures would have prevented it. Knowledge of producing toxic substances is valuable but does not guarantee regulators will act, assuming the production is somehow violating law. As Robert Jackson said in 1977, "I don't think having the information necessarily avoids another Kepone. I think it's what you do with the information that determines the effectiveness of this."[18] State regulators and local officials possessed the knowledge of Life Science's action, yet it took over a year to stop the company. There were legal tools available, but, rather than use them, officials demurred while the company continued to dump Kepone. Inside the plant, state regulators failed to share information and observations across agencies. Federal officials in OSHA, when made aware of the situation in the plant, mischaracterized it and failed to ask broader and deeper questions that might have brought swifter action. Aside from one cardiologist, doctors failed to grasp the possibility of Kepone poisoning. Earlier, at Allied, company officials in Hopewell were aware of their environmental requirements, later used against them in Judge Merhige's court, yet purposefully engaged in deceptive maneuvers.

Only a year later, in 1977, the state legislature returned to the drawing board to revise the TSIA. In between the original state act and the revisions came the federal TSCA, along with a host of issues that emerged from efforts to implement the 1976 law. Baliles led the revision in the House of Delegates, while J. Lewis Rawls Jr. of Suffolk led in the state senate. The main thrust of the original 1976 act was reporting and documenting the most dangerous of toxic substances and places where they were made and emitted, so that the state government could then monitor them. According to a memo prepared by Baliles, the original bill, when implemented, showed the "impracticability of publishing an all-inclusive list of all of those substances that might be toxic in varying conditions [and] concentrations." Baliles and others intended that a list "of those substances generally known to be highly toxic would be compiled and published." It was not intended "to require reports so voluminous as to be unnecessarily burdensome to industry"; such reports would "serve to obscure the essential information needed to carry out the purposes of the Act."[19]

Under the original act, industry complained at having a three-month turnaround to report on so many substances, regardless of degree of toxicity. Prior to the federal Toxic Substances Control Act or Virginia's law, industries generated hundreds of chemicals each year, exposing humans and animals in a largely unregulated system that benefited industry. At the time of these laws in 1976, some sixty-two thousand chemicals were already on the market, without assessments of their toxicity. When Virginia tried to create a list of some of these chemicals to then demand industry inform the government and therefore the public about their use, industry leaders complained the list was too long, the information on many substances unavailable, the expectations vague, the timeline for reporting too short (by January 1, 1977), and generally burdensome on their operations. Ironically, the same system that produced limited information on chemicals became a reason industry gave for opposing the 1976 Virginia law: they could not find information on the chemicals for which they were required to report. Faced with opposition to such a long list, the VDH revised the list down to four hundred and extended the deadline until July. Industry also continued to express concern that their reporting might divulge trade secrets on their substances.

Baliles led a task force that included state senator Rawls, members of the VDH (including Robert Jackson), and industry to amend the act, and the legislature approved those amendments, with support of several business leaders, in March 1977.[20] Amendments clarified that the state Board of Health would prepare a list of those substances that posed "the greatest immediate risk to public health" and dubbed these Class I substances. The board then held hearings on the list to gather information from experts and the public before finalizing it. Amendments also added that manufacturers had to notify the VDH "of those substances which reasonably should be known to be toxic under certain conditions" whether or not they appear on the list compiled by the VDH.[21] Manufacturers could also avoid filing duplicate information already required by TSCA. The VDH could also exempt many ancillary chemicals used in operations such as medical or water control labs, wastewater treatment, or building and grounds maintenance. The VDH also exempted agriculture and timbering, as they were not manufacturing establishments. In 1982 a new law eliminated the Class I designation, with the VDH arguing that the cost of evaluating each substance was expensive and that other legislation, including the Clean Water Act and the 1976 federal Resource Conservation and Recovery Act for hazardous waste, would be requiring listing of most hazardous materials anyway. Protecting industry trade secrets became a key part of the legislation. Only officials of the Health Department or agencies on the council had access to the toxic information. Federal authorities could access it if required by law and if they made a request in writing. The law mandated that any unauthorized disclosure of secret information subject the person disclosing it, or the person receiving it who makes use of it, "treble the actual damages sustained by the person whose secret information is disclosed."[22]

Even with the revisions, questions arose about the effectiveness of the TSIA. Newspaper articles from late 1982 in the *Richmond Times-Dispatch* charged that "state employees who should be using the collected information" were not doing so "because of fear of the possible consequences, i.e. treble damages." Only five officials of the State Water Control Board were said to have access to the confidential file held by the VDH. No one from the VDH's Occupational Health Department had access, unless they went through a "required procedure involving approval [by the health commissioner]" and the "assumption of liability for triple damages." As of 1982, no one at the Occupational Health Department had used that procedure.[23] The vagueness of terms like "confidential" and "authorized" also bred reluctance on the part of state employees to file reports against companies. A joint subcommittee of the General Assembly would conclude, "These articles also contained allegations that some companies were using the confidential file unnecessarily. Industry spokesmen maintained that protection of vital trade secrets was essential to maintain profits and employment levels."[24] On treble damages, the subcommittee clarified that those could only be assessed in cases of a "willful" violation. The subcommittee also recommended clarification on definitions of terms such as "authorized," "unauthorized," and "confidential." And it recommended protecting confidential information from the Freedom of Information Act, to avoid release of information should it be sent to other agencies. In March 1984, the Virginia legislature approved an amendment stating that state employees who mistakenly divulged confidential information would not face triple damages, only the prospect of being fired. Willful disclosure would be subject to those damages, along with a felony charge. Amendments also tightened the requirements for industry to prove secret formulas and other data needed to be confidential and allowed the health commissioner to share confidential information with physicians and firefighters in life-threatening cases.[25]

While there may have been questions about the effectiveness of the TSIA, the idea endured that citizens, first responders, and regulators need to know what is being manufactured, by whom, and where. The principle became part of later federal legislation, the Emergency Planning and Community Right-to-Know Act of 1986, intended to support emergency planning at the state and local levels for chemical accidents and to mandate that industry provide local governments and the public with information on chemical hazards in their community. This came about from the December 1984 Union Carbide accident in Bhopal, India, that killed some five thousand people after the release of methyl isocyanate, a chemical used in pesticides, and the release of the same chemical from a Union Carbide plant in West Virginia in August 1985 that injured 135 people living near the plant.[26]

Federal Toxic Substances Control Act

Just as it sparked new legislation in Virginia, Kepone featured in passage of the federal Toxic Substances Control Act in 1976. President Richard Nixon's Presidential Council on Environmental Quality originally proposed legislation to regulate toxic substances in 1971. Both the House and Senate approved bills in 1972 and 1973, but neither chamber could work out agreements over the extent and authority of the federal government over premarket testing of chemicals, the costs to companies, and the relation between the new law and existing ones.[27] A series of environmental contaminations including Kepone and "contamination of the Hudson River and other waterways by polychlorinated biphenyls (PCBs), the threat of stratospheric ozone depletion from chlorofluorocarbon (CFC) emissions, and contamination of agricultural produce by polybrominated biphenyls (PBBs) in the state of Michigan—together with more exact estimates of the costs of imposing toxic substances controls" helped ease opposition.[28]

The last and ultimately successful effort to approve legislation began in March 1975, just as Virginia regulators debated shutting down Life Science. Senator John V. Tunney (D-Cal.), the son of boxing champ Gene Tunney, sponsored the main Senate bill and chaired the proceedings. Not surprisingly, many chemical companies and the Manufacturing Chemists Association lobbied against it. One main argument was over the cost of testing. The chemical industry had net profits of

$5.5 billion in 1975, and the industry estimated the costs would be $350 million to $1.3 billion. The General Accounting Office estimates ranged from $100 to $200 million.[29] One industry spokesman said it would "cripple" scientific research and give "vast new powers" to the government. Others echoed this, saying the legislation would be too burdensome, likely to cause inflation or give government overly broad powers. One called it the "most serious threat the chemical industry has ever faced."[30] Despite the hyperbole, the bill passed the Senate with bipartisan support, 60–13, on March 26, 1976. Similar arguments emerged during the House hearings in July 1975, which opened just before Life Science and Kepone would become national news. There were three bills in front of the House Interstate and Foreign Commerce Subcommittee on Consumer Protection and Finance, introduced by Democrats William Brodhead of Michigan and Bob Eckhardt of Texas and Republican John McCollister of Nebraska.[31] After debates, the House approved a bipartisan new version sponsored by Eckhardt and Republican Jim Broyhill of North Carolina.

Kepone came up in the subsequent floor debates. In support of the legislation, the Senate Commerce Committee listed Kepone as the first example of dangerous toxic substances, along with vinyl chloride and PCBs.[32] Alabama Democrat James B. Allen, who chaired the hearings on Kepone in the Senate, referenced that experience in his arguments for reconciling language between regulations covering pesticides and those proposed for toxic substances. Republicans James Pearson and Lowell Weicker listed Kepone among other toxic substances as they outlined support for the bill. So did fellow Republican J. Glenn Beall of Maryland who noted: "In recent months, we in this country have become painfully aware of the catastrophic long-term effects that such chemicals as kepone, vinyl chloride, mercury, and PCB's may have on humans."[33] Congressman Brodhead cited the relationship between Life Science and Allied Chemical in offering an amendment that would have made it more difficult for large companies to hide behind smaller ones, which were exempt from many aspects of the proposed law. His amendment was defeated.[34] Robert Daniel, the Republican who represented Hopewell in the House, did not speak during debates but voted in favor of the legislation.

EPA administrator Russell Train also added his support. He and the agency came under criticism for failing to prevent Kepone from entering the James and PCBs from entering the Hudson. Three EPA lawyers resigned over what they charged as a failure of the agency to protect against toxic chemicals. Sensing an opportunity from the crisis and a chance to revive support, Train renewed his advocacy for toxic substances legislation. In a speech to the National Press Club in February 1976, Train stated, "[We] can no longer afford to wait for the basic authority to deal effectively with the problems of chemical pollution." He chastised industry for saying that the proposed bill would "cripple" chemical companies.

"The only real crippling that is going on," Train said, was that of "who knows how many Americans every year who contract cancer of some other affliction after exposure to some hazardous chemical agent."[35] After conferencing between the Senate and House, President Gerald Ford signed it into law on October 11, 1976, one week after Judge Merhige announced the fine against Allied.

Setting Kepone Action Levels

The debates over legislation to regulate toxic substances and the resulting acts revealed the tensions over the proper role of government and the rights and responsibilities of business. Those debates continued into the ongoing residual governance of setting and monitoring action levels of Kepone in marine life in the James River and Chesapeake Bay. Along with questions over the balance of power between government, business, and the public, the issue of Kepone action levels contributed to the growing discussions regarding the science of cancer related to toxic substances and what constitutes acceptable risk. Of course, the James River ban put science, politics, and economics on a collision course. Regulators had to respond to growing criticism from those in the seafood industry and their scientific and political allies. In the end, the action levels and the ban remained in effect through Republican and Democratic governors. Regulators did respond to new knowledge about Kepone levels and political pressure by revising action levels and lifting the ban on some species at different points between 1975 and the early 1980s. Eventually, in 1988, Governor Gerald Baliles allowed the remaining ban to expire as testing showed Kepone still in the marine system but below the action levels.

The Kepone task force coordinated the original response from Godwin's administration. To ascertain the scale and scope of Kepone residues, the task force supervised a monitoring program that included the State Water Control Board, Health Department, and VIMS. Through 1977, James Kenley chaired the group. That year Godwin reorganized it as the immediate crisis faded, and regular monitoring became its main function. William F. Gilley became the executive director in March. Godwin also added a Seafood Industry Advisory Council in September 1976, and its influence grew on altering the action levels on finfish in 1977 from .1 ppm to .3 ppm. The task force served as a window into how the knowledge of Kepone and its effects on marine life developed in the midst of the crisis, how that knowledge generated policy decisions, and how those decisions came into dispute.

The Kepone order issued by the Virginia Department of Health (VDH) on December 17, 1975, closed the river to fishing. The basis of the order was data from the federal "Preliminary Report on Kepone Levels Found in Environmental Samples from Hopewell, Virginia Area," published on December 16 that showed the

Virginia Marine Resources Commission (VMRC) inspectors check striped bass caught in gill nets in the James River. This photograph was taken in February 1985, three years before the end of the fishing ban on the James River. The contamination of marine life and the regular monitoring by state agencies like the VMRC were both residues of the Kepone disaster. Virginian-Pilot Archives/TCA

scale of contamination. The health order utilized the current research available on Kepone but noted that much was yet unknown about the pesticide. The order cited that ingesting "small amounts of Kepone may lead to accumulation and concentration in fat and body organs to levels which may be toxic to humans," and that according to the National Cancer Institute, Kepone was a carcinogen in test animals. The order also noted that "the public health effects due to the existence of low levels of Kepone in the environment are not known at this time," nor were the effects of "Kepone in the flesh of fish and oysters." Rather than keep the James open until more information became available, Godwin adopted a precautionary approach and closed the river first. The initial order prevented the "act of fishing, catching, netting, or taking of fish by any means" from the James and its tributaries from the Fall Line to the river's mouth. It did not mention crabbing, "mainly because we were under the impression that [crabbers] weren't taking crabs [from the river]," Kenley remarked. At this point, crabs had not been checked for Kepone. Kenley believed that since crab meat is low in fat, it would not be contaminated as quickly as fish or oysters. This would change by February 1976.[36]

At task force meetings in January, even before action levels had been set, Otis Brown urged that the task force set a date with the Health Department, April 1, for deciding on whether to open the James to fishing. No doubt he and Governor

Godwin were feeling pressure from the seafood industry, whose representatives pushed the state for information on testing of further samples from the James. But task force head Kenley continued to emphasize the need to determine a safe action level before harvesting fish. He continued to push the labs (in Virginia, North Carolina, Florida, and Maryland) to develop standards and add more testing. Godwin, despite criticism for closing the river, reiterated in public, "I do not think we can take chances with toxic substances of any kind." The task force remained open to new knowledge and adjusted regulations. For example, after testing by VIMS showed that seed oysters could depurate Kepone in clean water, the VDH opened the James for harvesting them for replanting.[37] Further investigations and testing revealed that shad did not bioaccumulate Kepone like other fish did, so the VDH later modified the original to allow for shad fishing.[38]

The Kepone action level proved the key factor in the fishing ban. Setting these levels came from a cooperative arrangement between the EPA and FDA, based on laws establishing each agency's authorization. The EPA studied pesticides and then could set limits, if any, on exposure to Kepone. The FDA had the power to regulate substances in food, so that agency issued the directive and had responsibility for enforcing any bans on shipment or sales. In setting the action level, these agencies were asserting a level of exposure below which the body can adjust without causing harm. They were following modern toxicology which "normalized the problem of low-level chemical exposures" in a world rife with toxic substances.[39] By March 1976, the agencies set the levels for finfish and eels at .1 ppm and oysters and other shellfish at .3 ppm. By late March the agencies set .4 ppm as the action level for crabs.

The basic procedure for setting levels in food used a "no-effect level" of the substance, the weight of a person, and the amount of fish or shellfish someone would eat in a year, using per capita consumption in the United States. The method certainly wasn't foolproof and opened the door to challenges. There was uncertainty in other aspects as well. The EPA noted that Allied Chemical suggested a no-effect level of 1 ppm for humans, based on the company's earlier studies of rats. Yet other studies showed ill health effects below this, so the EPA approached this number cautiously. In addition, at the 1 ppm level, tests showed kidney damage. The weight of the average person was sixty kilograms, or about 132 pounds, and per capita consumption came from national figures of marine species. The agency then set safety factors to account for sources of uncertainty, according to the following conditions. The EPA applied a factor of 10 to account for the subject population that "includes not only healthy persons but also aged, ill, and infants." Then came an additional factor of 10 to account for the "range of susceptibility among test animal species (10 x 10 = 100)." Last, another safety factor of 10 applies when there is a "case of a pattern of severe effects with long-term implications" that include effects on reproduction or genetics causing cancer. So 10 x 100 = 1,000. Usually the EPA adopts the 100 factor, but the 1,000 factor was not unusual "where the totality of effects" required it. This would mean that an action level for a compound would be $1/100$ or $1/1000$ of the amount that causes no observable adverse effects.[40] In addition, there is a decision by scientists regarding the action level and then one at the higher level of policy making in the EPA. These decisions involve considering the potential of a substance to cause cancer, the sensitivity of the methods in measuring the smallest amount for setting the level, and then economic impact on the food in question. It seems the scientists "originally recommended an action level of zero for Kepone," but only the economic impact moved the EPA policy makers to apply the thousandfold safety factor (as opposed to the usual hundredfold) to reach the issued action levels.[41] In doing so, regulators established this threshold model for exposure, essentially drawing an arbitrary line in carcinogenic space to accommodate the fishing industry to mitigate against an economic collapse.[42] From all this, the EPA concluded that based on the "demonstrated carcinogenic potential of Kepone in two species of test animals, the lack of a clear cut 'no-effect level' in the two year rat feeding study, and the evidence that Kepone is a cumulative toxin," the EPA established and maintained the original action levels of .1 ppm for finfish, .3 ppm for shellfish, and .4 ppm for crabs.[43]

Concern in 1976 and 1977 remained high among many scientists studying Kepone, reflecting fear and uncertainty. In the first of two Kepone seminars sponsored by the EPA, Bob Huggett of VIMS claimed that Kepone was likely one "of the most chronically toxic substances" humans had to address. They remained worried over how Kepone easily entered the food chain and how to dispose of the waste and address the toxic sediments in the river. They feared the entire Chesapeake Bay might be ruined.[44] And while scientists worked to gather more data on Kepone's toxicity, others showed how certain fish and oysters could depurate the poison. This gave some hope that the initial fears of a massive toxic event might subside. Perhaps the river wasn't hit as hard as it initially seemed. Throughout 1976 and into 1977, the Health Department remained flexible and adjusted the order as new evidence emerged. For example, like shad and oysters, channel catfish could depurate Kepone to limit the amount of the pesticide in the edible part of their bodies. Research also showed that male and female crabs showed differing levels of Kepone. Female crabs stayed closer to the salt water of the Chesapeake Bay to spawn, whereas male crabs migrated farther upstream, where they could be exposed to more Kepone. Over time, the VDH opened the river to harvest of all catfish as well as elvers (baby eel), as migrating elvers do not possess a digestive system for four to five weeks after entering the James River, and after being caught they are raised in Kepone-free ponds.[45]

Yet as these findings opened some species for harvesting, at its meeting in November 1977 the Health department recommended closure for harvest of other key species until 1980. Kenley stated that having six-month increments "had tended to hold out false hopes for many of those involved in James River fishing."[46] Godwin did not authorize that, but he did extend the ban on January 1, 1978, for one year, allowing for midyear revisions if needed. When Republican governor John Dalton took office later than month, he continued the ban as recommended by Kenley and the Health Department, which lasted until 1980—as Kenley had requested in 1977. Kepone levels remained higher than the action level for finfish and crabs.[47]

Debating the Ban

As Otis Brown remembers of the ban, "We really just decimated the oyster and seafood industry." Understandably, watermen were upset. "'The bib-overalled people,' we called them, because they'd just come meet me with their bib overalls. There wasn't any cleaning up to come to the state capital. They were a tough bunch of people." "But," he notes, "on the other hand, they understood. What they didn't like was the fact that this was something new, and no one really knew what the impact was." The eel fishermen were even tougher, according to Brown. When shutting down the eel industry, Brown remembers, "I had a plain clothes state trooper go with me [to meet with the eel representatives]. They were not a happy bunch."[48] Chemist Bob Huggett of VIMS concurred. "We had some very scared people," he recalled. They were angry too. Some watermen "launched bomb threats and death threats against VIMS researchers." Huggett slept with a shotgun by his bed.[49]

It was clear why they were all upset. Sales had dropped as a result of the Kepone contamination, and now they were faced with closure. Overall, the James River produced some 14 percent of the total value of seafood from Virginia intended for human consumption. Yet more than Kepone was at work. Of finfish in the James, rockfish or stripers held the most value, but they were in crisis by 1975, with landings in overall decline.[50] The James was especially important for seed oysters as well as blue crabs. One estimate calculated that 75 percent of seed oysters planted in Virginia waters came from the James.[51] Yet landings had been in decline since the early 1960s. Although the value for blue crabs was going up, dockside landings for blue crabs had declined since the 1960s as well.[52] Crabbing was of particular concern along the peninsula of the lower James. Loss of crabbing would account for rates of short-term unemployment for some 60–70 percent of watermen in some areas of the peninsula. In Hampton, loss of crabbing might not affect many watermen, but with a crab processing plant the impact could be significant.[53] Findings of Kepone-contaminated bluefish in the Chesapeake in the summer of 1976 added to the growing concerns over the state's seafood and also threatened the closure of the entire Chesapeake Bay, as did estimates from VIMS and the State Water Control Board that anywhere between ten thousand and one hundred thousand pounds of Kepone remained in the sediments of the James.[54]

Joan Youngblood, associate executive secretary of the Virginia Seafood Council, outlined some of these issues during the Senate hearings on Kepone in January 1976. She reminded those in attendance of the economic impact of closing the James: "1,453 individuals in the area immediately adjacent to the James," she said, earned all or part of their income from the river, with another 1,273 involved in processing seafood. Seed oysters amounted to some $1.1 million in dockside value, and finfish around $410,000, with retail values three to four times higher. She echoed the view of many crabbers regarding regulation and pollution. "We do believe, 'Let industry come' but not regardless. Hopefully, this event will be an example to the regulatory people and legislators at the State and Federal level to provide the money and the expertise so that such an event will never occur again." Weston Conley Jr. of the Virginia Oyster Packers Association noted that the entire oyster industry in Virginia employed some seventy-five hundred people and was worth about $100 million each year. He added that his group had been active in getting the test results that showed oysters could depurate Kepone. Conley was less than supportive of the state regulators and setting action limits. "In all fairness to our State agencies, I do not think we have ever had a situation concerning our environment that has ever caused such mass confusion."[55]

Crabbers, who had been exempt from the initial ban in December 1975, felt the pinch after Virginia imposed one by February 1976, with a focus on male crabs. The state opened the harvesting of females in the spring in the lower James, as that area sees an annual in-migration of uncontaminated crabs from the ocean. As they continued their lawsuit against Allied, crabbers also spoke out against the action levels and the ban itself, using both an economic argument as well as one resisting state knowledge over scientific data. "It hurts a lot when they take away the place you earn a living," said James Wilbur Firth Jr. He lived in Poquoson and had eighteen crabbers working for him. Angry that Kepone took away his livelihood without compensation, Firth said he was concerned: "Young people ain't going to take to the business." A Portsmouth crabber, Ernie Russell, disputed the state's testing data. "There ain't nothing wrong with any of them damn crabs. I've had them tested at my own expense and they tested below the action level." He did not elaborate or provide details of these tests. Russell added that the Health Department had "acted irresponsibly" over the preceding two or three years.[56] Along with banning areas for crabbing and restricting harvest of male crabs altogether, in 1978 the state went so far as to seize and destroy some crabmeat on the market, and two crabbers were convicted of harvesting male crabs and fined $100, with a six-month suspended sentence.⁵⁷ Later, in 1979, as Kepone levels eased, male crabs could be harvested in parts of the Lynnhaven River and Lynnhaven Bay.⁵⁸

Kepone illustrated how watermen whose lives depended on crabs and oysters were often at odds with regulators in Virginia and the federal government. As Michael Paolisso notes in his work on crabbers, many believe God's stewardship of nature provides balance and regulation. They are dubious of the ability of science to predict crab populations. "Time and again watermen said scientists are smart people and know a lot, but you cannot know the blue crab unless you work the water and try to catch the crab year in and year out. It is only through experiential knowledge that one can learn what can be learned about the blue crab."⁵⁹ Indeed, as Christine Keiner notes about Maryland watermen, they had a "fine-tuned knowledge of the local biotic conditions" that required an "intimate relationship with the local oyster commons."⁶⁰ They generally resist regulations on harvesting but support science in studying water quality and support regulations that reduce the harmful effects of pollution. With Kepone, their concern was over the damage caused by polluting the river, yet they fought the ban on harvesting. Anglers focused on finfish expressed similar concerns. Shad fisherman Thomas Hazelwood of Suffolk told the Senate subcommittee during Kepone hearings, "I am a full-time waterman as has been traditional in my family for 100 years on the James River." He went on to note that he and his fellow watermen had "seen a decline in production in our industry [and] feared that chemical pollution was part of the problem" "Our river is used as a cesspool by some cities along its bank," he added, "[but] nothing ever seems to be done by the bureaucratic governmental agencies and processes in Richmond [who] have failed us in a duty which is their sole purpose for being." "We sincerely hope that our river can be cleaned up so that we and our children can build our industry back up to what it has been in the past," he said. William Gay, oyster tonger, added, "We only want our God-given right to work that river returned to us." "If we have to wait for the mechanics of all this testing," said Fred Garrett of the Virginia Oyster Packers Association, "it will be way beyond the fact and a lot of us will have tremors from poverty and not much from Kepone."⁶¹

The Kepone crisis fed into a simultaneous one over the actions of VIMS director William J. Hargis. Hargis's agency played a key role in testing Kepone and in locating the frozen seafood that proved Allied had dumped Kepone in the river before Life Science had and put a black eye on the State Water Control Board. He also expressed initial concern about closing the entire Chesapeake Bay to harvest of bluefish, angering an already hostile seafood industry. Seafood processor Charles Amory Jr. of Hampton said, "[Hargis] has been trying to get the Bay closed since January. He's looking for more money and more personnel, and he's using this Kepone thing to drum up support for VIMS."[62] Hargis—along with the seafood industry and others—also came out against one of Godwin's projects: a new oil refinery in Portsmouth along the Elizabeth River. Hargis had also used deficit financing to spur growth of VIMS, leading to concern among some state leaders. Amid these tensions between Hargis and the seafood industry, Hargis found himself investigated by the state police. He was arrested and charged with attempted embezzlement and wrongfully disposing of state property—an unused diesel engine from a VIMS research vessel. Asked by reporters at the time if the charges "might be the result of efforts by influential interests displeased with VIMS' findings on Kepone," Hargis replied: "I'm sure there has been a number of people who are unhappy with things we've had to say from the institute, based on data—but I don't believe there is such an effort."[63] The charges were dismissed, and Hargis stayed on at VIMS until 1981.

In its continued effort against the James River ban, the seafood industry enlisted the help of Virginia state senator Herb Bateman, who led the resistance to action levels using arguments that dovetailed with those of crabbers over whose knowledge could be trusted. Godwin had tapped Bateman to shepherd passage of the Kepone package of bills in the Senate, and Bateman did so. But before long the senator turned toward dismissing Kepone's toxicity and attacking the action levels. He represented the peninsula area where the seafood industry had influence and economic impact. Over time, Bateman was not alone in pushing to raise the action levels. Godwin also began advocating this by the summer of 1976. James Kenley's views also evolved from holding firm on action levels to pushing for an increase.

Another key advocate against the levels was James Douglas, head of the Marine Resources Commission. As Howard Ernst notes, the commission "evolved from a fisheries commission that was created to protect and enhance the state's commercial fishing industry, not to pursue broad ecological objectives."[64] The commission tends to support watermen over conservationists and scientists, often emphasizing "economic primacy" in environmental policy related to Virginia waterways and the Chesapeake Bay. In the Kepone case, Douglas began correspondence with Bateman as early as August 1976 and pressured Kenley too. He relayed concerns from other Virginia Peninsula political leaders opposed to the action levels, including Lewis A. McMurran Jr. and Theodore V. "Ted" Morrison of Newport News and Hampton's Richard "Dick" Bagley Sr. At the behest of seafood industry leaders, Bagley urged Douglas to argue against having the state agencies be "the sole source of review" of data and that there be a "blue ribbon panel of scientists" outside the state's Kepone task force to review the EPA's recommendations. Douglas also questioned whether the testing used by the EPA was, in fact, finding Kepone and not some other substance. He referenced a story about how a scientist looking for DDT accumulations discovered substances later identified as PCBs.[65] He even recommended purposely trying to "fool" labs to make sure they were accurately identifying Kepone.[66]

The Virginia Seafood Council and National Fisheries Institute prepared their own "Kepone Fact Analysis" to counter negative publicity over Kepone. They argued that Kepone contamination had been "exaggerated." They noted that with the thousandfold safety factor used (as opposed to the standard hundredfold), the average person would need to consume twenty-four thousand pounds of fish per year to be exposed to the 1 ppm amount that caused harm to test animals. Consumers should just be sure to consume a wide variety of foods, including fish, so as to avoid illness.[67] They also prepared a marketing campaign aimed at restaurants and "the housewife," noting in an internal document that prior "to the Kepone problem, no other environmental condition which could be considered detrimental to the quality or wholesomeness of Virginia seafood had received public attention." They proposed the get consumers (mainly women, perhaps) to "forget" Kepone in the same way they had forgotten about the fear of mercury in tuna in 1970 or aminotriazole in cranberries in 1959. They placed ads in daytime television shows, newspapers, and magazines touting the benefits of Virginia seafood.

Bateman initially commended Godwin for doing what was needed to protect public health, but he became more critical as the ban went on, urging the state not to "overreact to the health hazard." He urged Virginia officials to question or seek others to question the FDA and EPA action levels. He noted that some authorities with whom he had spoken raised doubts about the criteria used to set the action levels and the "validity of the assertion that Kepone is a carcinogenic agent." Bateman urged that the seafood industry be allowed to join in deciding on testing and action levels. He wanted more positive news about Virginia seafood and a controlled statement from only the governor or one or two designated officials, seeing Kepone as casting an "unnecessary cloud on the marketability of all Virginia seafood products."[68] He remained skeptical that Kepone was carcinogenic, stating in August 1976 that of the employees directly affected by Kepone at Life Science, "whatever other health problems they may have, none have been found to have cancer."[69]

To help his cause, Bateman consulted experts outside government. Of those, he came to rely on scientists who favored continued use and development of chemicals and pesticides, who held a critical view of regulating industry in general, and who questioned the science behind those regulations. Whether consciously or not, in doing so Bateman was part of a larger movement in the 1970s of corporations, trade association, scientists, and politicians pushing back against the flourishing research into and regulation of environmental pollutants. These groups challenged findings on a range of issues, from acid rain and the hole in the ozone layer to cancer-causing agents like tobacco and various chemicals. Kepone became part of these battles, especially those focused on the environmental causes of cancer.[70] Typical arguments enlisted in the defense of other chemicals often mentioned Kepone and said that cancer is more a product of lifestyle choice than industrial toxicity or occupational exposure, that there is a safe, low-level exposure for toxic substances (and everything can cause cancer if you consume enough of it), that the use of chemicals in protecting society outweighs the risk (rare are chemicals that are truly dangerous), that fear of chemicals is worse than any chemical itself, that evidence of toxicity is inconclusive or incomplete, and that more research is needed.[71]

One expert upon whom Bateman came to rely was William Deichmann. Deichmann (1902–90) was born in Kiel, Germany, and naturalized in the United States in 1930. He earned his PhD from the University of Cincinnati in 1939 and became chair of the department of Pharmacology at the University of Miami in Florida. Active in the field of toxicology, especially in pesticides and occupational health, he had a long record of consulting with industry, including clients such as Pfizer, Dow-Corning, Monsanto, Velsicol Chemical, and Toms River Chemical.[72] He received the Cummings Award from the American Industrial Hygiene Association in 1975. His lecture for the occasion, "The Market Basket: Food for Thought," fell in line with other proponents of pesticides and other chemicals as necessary for global food production. He insisted on the adage that "the dose makes the poison," that only the dose determines whether a compound is toxic to humans. He did not address the effects of these chemicals on animals other than humans. Merely retaining chemicals in one's organs or tissues did not provide evidence of toxicity, per se. He accused the media and other "alarmists" of "playing on the emotions" with "exaggerated accounts of poisonous chemicals in our food products." He cited the DDT ban as one example of weak arguments on the health effects of the compound on humans, noting that the evidence on DDT as a cancer-causing agent in humans was inconclusive. Deichmann was among those who studied the toxicity of DDT, coauthoring a study sponsored by Geigy, the originators of DDT, which downplayed the toxicity of the chemical in continued use in agriculture.[73] He also criticized bans on Aldrin and Dieldrin for similar reasons, saying that the studies showing their cancer-causing properties on humans were only done in mice. He preferred more evidence from human exposures to truly assess the dangers, citing evidence of occupational exposure to both chemicals by workers in the Netherlands that revealed there was a clear no-effect level.[74] It is not clear how scientists were to gather these human experiments, other than waiting for another accident or deliberate experimentation.

It is also not clear how Bateman and Deichmann became connected, but in September 1976 Bateman asked him to prepare a review of Kepone and the action levels. Deichmann provided counsel as Bateman prepared his attack on action levels, urging the state senator to obtain all the information he could, especially on the concentrations of Kepone in the men who were poisoned, and to consider "the economic importance of Kepone"—both points Bateman underlined in his notes.[75] The focus on economic arguments underscores the shift in the 1970s to "quantitative risk assessment," whereby risk is assumed to be unavoidable, managed rather than eliminated. Regulators then manage poisons based on allowing a minimum standard of exposure before harm is likely to occur, "a trade-off for economic gains."[76] Of course, setting action levels is part of this method of risk assessment as well, as the EPA took economics into account in its decisions.

In his November 1976 report, Deichmann acknowledged that while high doses in test animals induced negative effects, he noted that the "data presented offer[ed] little concrete information on low dietary concentrations of Kepone that might be considered harmful or harmless for man." He argued that tests done by the NCI could only be considered a "rough screening for carcinogenicity." He questioned whether test animals could have been exposed to other environmental chemicals besides Kepone that may have caused their carcinomas. He also noted how tests showed concentrations of Kepone in the liver, which was something Guzellian had found with workers, and that, once Kepone was removed from the diet, residues in test animals decreased. In one study led by Paul Larson in 1960, dogs and rats fed Kepone showed a no-effect level at 1 ppm; Deichmann concluded that this was the accepted no-effect level for Kepone (a finding Bateman underlined in his copy of the report).

Deichmann went on to criticize the FDA for using national intake of fish and shellfish as opposed to local or regional numbers for determining the action level. He argued that the average level of Kepone in finfish for the whole Chesapeake Bay and James River was only .03 ppm (which ignored the higher levels in the James). This meant that most consumers of seafood would only ingest .000345 mg of Kepone per day, well below the action level set by the FDA/EPA. Given this, assuming the standard safety factor of 100 or even 160, Deichmann felt confident that the action level could be increased from .1 ppm to 16 ppm—based on what most consumers would eat. "But to be conservative, one could with reason recommend that the permissible concentration of Kepone in finfish be increased from .1 to 2.67 ppm," in line with heavy fish eaters. He went further and concluded that since rodents with high concentrations of Kepone did contract liver cancer, he recommended a permissible level of .5 ppm in finfish.[77]

Bateman also received advice from Thomas Truitt of the law firm Wald, Harkrader & Ross. Truitt was active in representing Occidental Petroleum in its litigation over Love Canal. The firm also worked on raising limits on PCBs in food packaging and was active in the Food Safety Council, a business and pharmaceutical trade association to combat consumer fears of food additives and chemicals.[78] As he went about his work as a legislator, Bateman also earned money as a lawyer representing the Virginia Seafood Council. They paid him from its SOS Fund—created through donations to aid the industry in addressing the Kepone crisis. Bateman helped create the fund.

Bateman arranged a meeting among members of the Virginia Seafood Council, the National Fisheries Institute, and the EPA in September 1976. Their concern was finfish. Leaving alone the technical basis of the action level, industry challenged using national data on finfish consumption. Instead, the seafood industry argued for using total commercial consumption from the Chesapeake Bay only, which would raise the action level from .1 ppm to .3 ppm, allowing another million pounds of product into the market.[79] Bateman pushed for a quick answer, continued to work through Otis Brown to have the Godwin administration join in pressuring the EPA to modify its action level, and lobbied Gerald Ford's administration.

The EPA responded to the pressure and conducted hearings in Baltimore and Richmond in late January 1977. Bateman made his case in Richmond for the seafood industry as their paid spokesperson, reinforcing the views of Deichmann and the quantitative rationale of risk. The original action levels had been set "in an atmosphere of crisis and without all the data which would have been desirable," he offered. He argued that "a precedent of over abundance of caution" on Kepone and other chlorinated hydrocarbons could lead to "a practical ban on most foods consumed by man."[80] A zero tolerance for these would lead to widespread starvation among the peoples of the world. Data, he argued, supported "a balanced judgement which accepts continued availability and consumption of residues of unavoidable chlorinated hydrocarbons in our food." The "real world" we live in, he went on, "is not an environmentally pure one." There is a wide variety of contaminants, and risk is unavoidable and must be weighed against the benefits. He compared Kepone and other chemical residues to the deaths attributed to automobile and airplane crashes. "No one suggests ... that we ban motor vehicles or planes. We accept risk from these sources because the total risk benefit analysis dictates that we do so." Essentially, regulation and action levels came down to the issue of how safe is safe enough. Bateman noted that the action level of other pesticides like Aldrin, Dieldrin, and chlordane in fish was higher than Kepone, even though those chemicals were acutely more toxic. He followed the report of Deichmann in questioning the efficacy of the NCI tests and sowed doubt into extrapolating testing in animals to effects on humans. He then focused on the cancer risk. He dismissed evidence of a cancer-pesticide connection by stating that there was "no single known instance where the residues of any chlorinated hydrocarbon [had] resulted in or caused cancer in any human."[81] Of course, the animal studies suggested a strong connection, and there were at that time few examples of human exposure to chlorinated hydrocarbons, aside from the workers at Life Science.

By this time, James Kenley of the Kepone task force had swung his support to raising the action level. At the Richmond hearing, Kenley stated that the overriding concern remained "the health of our citizens": "No change should be suggested which places the population at any significant risk. If, however, modifications can be made which will allow us to stay within those safety limits and still allow for improvements in the economic position of our state's watermen and the seafood industry, then we support that concept."[82] Robert Jackson, the man most responsible for exposing Life Science said, he supported "moderate increases in the action levels" but only so long as warnings about limiting consumption remained. The Allied spokesperson at the hearing argued that the studies done by the NCI that formed a key part of the research on setting the levels had little scientific validity and created a "cancer scare."[83] Dr. Mike Bender of VIMS chaired the subcommittee of the task force examining the monitoring data from the James. He concurred that the action level could be increased. His report to the EPA reviewed the extensive sampling program conducted by the state and the knowledge gained regarding how oysters could depurate Kepone when removed to clean water and that male crabs spend greater time upriver in the fresher water of the James than do female crabs. He noted that the longer a particular fish spends in the river, the more Kepone is found in its body. Fish in the Chesapeake Bay that were closer to the James, along the western side, also had greater Kepone than those sampled from the eastern side. Bender showed that the original action levels also took into account safety factors to include foods other than seafood that might contain Kepone. Testing showed that was unnecessary as there was no Kepone in other foods. He also agreed with seafood industry's argument that the baseline average intake for seafood of residents was too high. Based on this, Bender argued for action levels of .3 ppm for both finfish and shellfish, and .5 ppm for crabs.[84]

The pressure and the presentation did have a limited effect on the EPA and RCRA. The agencies agreed to raise the action level in finfish to .3 ppm but maintained their reliance on the thousandsfold safety factor. While this provided minimal satisfaction to the seafood industry, the decision prompted the Environmental Defense Fund (EDF), an advocacy group formed in 1967 over the presence of DDT in osprey eggs on Long Island, to go public on Kepone and oppose the decision. Dr. Joseph Highland, chair of the EDF's Toxic Chemicals Program, argued before 1977 Senate subcommittee hearings on worker safety in pesticide production that the decision completely disregarded "the scientific accepted fact that there is no safe level for a carcinogen." In raising the level, the agency ignored internal documentation from scientists estimating the increased risk of cancer from consumption of fish contaminated with Kepone at different action levels. Moreover, as the EPA recognized, there was no "threshold level" or no-effect level for Kepone's carcinogenic potential, and Kepone was a "cumulative toxin." In addition, while Deichmann and the seafood industry wanted to dilute the amount of seafood presumably consumed to raise the action level, Highland argued the opposite. Just as setting an action level should "represent an upper limit of the amount of the chemical which is actually found in the food, so too the amount of fish consumed should represent an upper limit" in order to protect "the total population, not just those people on the lower end of the consuming spectrum." With so many unknowns and Kepone's carcinogenic makeup, Highland argued for action levels at the level of detection by then-current testing methods, which was 0.02 ppm, far below the levels of .1 to .4 ppm, depending on "further analysis of risk and economic impact."[85]

As Highland's testimony indicated, the economic impact seems to have been the deciding factor for the EPA in recommending action levels. Although, in making that assessment, the EPA's reading of the socioeconomic impact of the ban still did not necessarily favor those along the James. Edwin Johnson, deputy assistant administrator for pesticide programs at the EPA indicated that he made the decision to raise the action level in finfish to .3 ppm in order to open the Chesapeake Bay but still keep the James mostly closed.[86] Along with helping the seafood industry gain a slight modification in the action level, just before Godwin left office he earmarked $500,000 of the legal settlement from Allied for the marketing of Virginia seafood, which became the Virginia Marine Products Commission.

Ending the Ban

By 1980, Kepone fatigue was setting in. Violations of the ban had become more common. Even a member of the Marine Resources Commission—the agency then enforcing the ban—was cited.[87] Watermen would be settling their lawsuits by summer. In a shoot across the bow of state regulators, in May Newport News circuit court judge Henry D. Garnett overturned the Kepone ban on the James River. According to the judge, the Health Department was thirty-four days late in submitting for renewal the emergency order under the Administrative Process Act. Garnett did not consider the medical evidence on Kepone in his ruling. Lead counsels in arguing to overturn the ban were Hardaway Marks of Hopewell and Theodore Morrison of Newport News. Some suspected pure politics in the decision: they were also the second- and third-ranking members of the House Courts of Justice Committee, with great influence on judge appointment in Virginia. The Health Department asked the Virginia Supreme Court for a stay of the order, but the court did not immediately act. Republican governor Dalton supported the Health Department and used his executive power to reinstate the ban. The ban may have been helping ease consumer fears over Virginia seafood. Prominent Virginia waterman Cranston Morgan thought the levels could be increased. Yet he praised Dalton for continuing the ban, as the fear of Kepone would have kept most Virginia seafood out of the marketplace as it had done in 1976. With the James River catch "but a grain of sand on the seashore compared with the Virginia total catch," any delay in the ban would have been disastrous for the industry generally. While potentially devastating for the James River industry, those outside the James were doing well in the wake of the initial fears of the entire Virginia industry being damaged beyond repair.[88]

With fish continuing to show levels of Kepone higher than the action level, in 1980 the Health Department considered a five-year ban as opposed to six-month or yearly ones. So even as Kenley and others sought to raise the action level, the rest of the Board of Health maintained the ban as long as evidence showed the higher Kepone levels. A public hearing on the matter prompted an outpouring of resistance from Bateman and others advocating for the seafood industry. Bateman maintained his opposition. He insisted that higher action levels were safe (or could be removed entirely) since consumers would never eat enough seafood in their diet to reach the highest end of exposure. This was, he claimed, "a classic emotional response to a problem." He asserted later that officials in the EPA saw the case for raising the action levels above those established, but that since the state did not push to go higher, the EPA did not either. He continued to insist that the testing and methodology for setting action levels was flawed, that humans can depurate Kepone on their own, and that the state should consult toxicologists outside the government, toxicologists whose work Bateman used as basis for his arguments.[89] Hardaway Marks had voted in favor of Godwin's Kepone package in 1976, but he called the lawsuits stemming from Kepone "the biggest fraud ever concentrated or pulled upon this Commonwealth" and argued that he had learned to swim in the James River "when there were no sewage treatment plants" in Petersburg or Richmond, and that neither he nor others have suffered any ill health. "It's been too long for this Commonwealth to have this fraud continue," Marks said. "[It's] time for this river to be opened up." Watermen denounced the continued ban, agreeing with Bateman that the process was flawed and that the state needed to involve them more in sampling and testing. The loudest cheers came when Hampton waterman Sonny Ballard said, "I feel that the pollution's in Richmond, not in the river."[90]

By this point, Kenley seemed to be inclined to alter the ban in favor of fishing. But he was hesitant, worried that new publicity would again be injurious to the seafood industry. Douglas assured him that hearings would not be "anywhere near as damaging as . . . back in 1976." Douglas was confident that "additional studies and continued monitoring data bring us only good news!" Further, if studies that "reveal the human's capability to metabolize" Kepone come out, along with those that "cast doubt on the labeling" of Kepone as carcinogenic, then these would only increase consumer confidence.[91] After the July hearings, with Kepone levels lower than years before, Kenley recommended to the Board of Health that the original plan for a five-year ban be reduced to eighteen months and that the river be opened for all commercial and recreational fishing from January to June, when historically Kepone levels were lower. Yet the board was not ready to go this far immediately.[92] In September, the Health Board voted to allow recreational fishing and consumption of fish, albeit with warnings attached.

Then, after another emotional hearing in November 1980, the Health Board voted to eliminate significant parts of the ban on commercial fishing. Virginia allowed fishing for all finfish other than striped bass and eels during the first six months of the year, before Kepone accumulation reached action levels. The new regulation also opened the lower James for commercial harvesting of male crabs from January through September. In the watermen's favor was that 1980 was a drought year. With river flow lower than average, there was less Kepone moving in the water. Crabbers were especially jubilant. "This is a tremendous breakthrough," declared Roy Insley of the Virginia Working Watermen's Association. He called the Kepone threat "largely a hoax." "I don't know of anyone who has suffered from Kepone except watermen." He apparently forgot about the poisoned workers. The data on crabs was a particular point of contention. The watermen produced their own studies showing that it was safe to extend the harvest into September, whereas data from the Health Board's staff recommended ending it in June. Some male crabs still showed Kepone levels of .5 to 1.3 ppm, above the .4 ppm action level. The board sided with the watermen. One board member, Dr. Virgil H. Marshall, likened the new system of posting warnings on Kepone to "the health warning on cigarette packages." These statements referencing tobacco connect how eating seafood potentially contaminated with dangerous chemicals had become what Robert Proctor refers to as a "lifestyle cancer" like smoking. Individuals were seen as having choices; there was no need to look further at larger sociopolitical contexts of tobacco or chemicals.[93] James Douglas, head of the Virginia Marine Resources Commission, pushed to end the ban entirely, arguing that the action level for Kepone was unrealistic. "It comes to a question of whether we have at these levels a threat to human health," he said. He attacked the federal bureaucracy, saying, "You can't see its ears or its mouth, and when you want to kick it you don't know where." Theodore Morrison argued that the entire Kepone ban had become a joke to many. "Everybody's violating it and I say God bless them."[94]

As he slowly moved toward easing the ban, Kenley continued to push the EPA to raise the action levels. He based his reasoning on the earlier studies done by Guzelian showing that Life Science workers' bodies metabolized the pesticide as Ke-

An angler waits for a bite along the James River in September 1976, apparently unconcerned about the warning sign next to him. The State of Virginia issued a fishing ban for the James River that lasted from 1975 to 1988. The ban proved controversial and was not highly enforced against individuals. The seafood industry fought hard against it, and eventually the government removed the restrictions, even though Kepone and other toxic substances continue to be found in marine life. Virginian-Pilot Archives/TCA

pone alcohol, which some contended did not cause cancer. The theory was that since rats did not metabolize Kepone in this way, studies comparing carcinogenicity in rats to humans might be invalid. Kenley implored the National Institute of Environmental Health Sciences to resolve the issue of Kepone as a cancer-causing agent. "I cannot emphasize too greatly the importance of this question to the citizens of Virginia."[95] The arguments did not sway John Moore, deputy director of the National Toxicology Program. He argued that both rats and humans absorbed ingested Kepone readily into the gastrointestinal tract and distributed it to tissues. They both excreted Kepone slowly, and the pesticide stayed longer in humans than it did in rats. He also noted that some agents might not be genotoxic but are carcinogens.[96] A genotoxic carcinogen "indicates a chemical capable of producing cancer by directly altering the genetic material of target cells," while non-genotoxic chemicals cause cancer by secondary means not related to gene damage, which could include disrupting cell structures, the rate of cell proliferation, or hormonal changes."[97] Often, the latter category (like Kepone) have threshold levels associated with them, and for some the effects can be reversed when exposure ends. In addition, organochlorides like Kepone often induce liver tumors. Some are also endocrine disruptors affecting reproduction and development. Genotoxic chemicals have no threshold level. Some chemicals are difficult to separate into either group. Moore wished to avoid excluding Kepone as a possible carcinogen and raising the action level any higher. William Dykstra of the EPA also noted that new data had emerged by 1980 that showed estrogenic effects in rats, birds, and mice, and oncogenesis at 1 ppm in rodents. In addition, some tests showed that Kepone did exhibit mutagenic properties like genotoxic materials that had no threshold level. This meant that there was no acceptable level of exposure for Kepone. Moreover, Kepone showed a greater effect on females than males in tests, and no data existed on how Kepone might affect women. In fact, tests done in the 1980s showed a higher incidence of cancer in female rats than male ones. Given this latter set of evidence, the action level remained.[98]

Later arguments from public hearings in March 1982 on the continued ban on commercial fishing echoed earlier ones, despite the caution exhibited by Moore and others. James T. Mathews of the Working Watermen's Association pushed to end the ban on commercial fishing. He argued that eating fish from the James since the 1960s hadn't caused him or other watermen any observable health problems. Other sources of pollution were of greater concern than Kepone, such as sewage. John DeMaria of Newport News, former president of the Watermen's Association, agreed. He criticized the levels as being too low. Like others, he used the analogy of the warning label on cigarettes as being sufficient to protect public health. A Kepone warning was sufficient along with open fishing—amid the heavy marketing blitz assuring consumers that Virginia seafood was safe. Similar debates continued through the mid-1980s. Only in 1988, with Kepone levels for all tested species below the federal action levels, did Governor Gerald Baliles recommended to the Virginia Board of Health that the state lift the last ban on commercial fishing, which they did. This ended restrictions on stripers, croaker, and eels, and brought to a close the often-bitter, thirteen-year controversy created by the restrictions.[99]

Mike Unger is a professor specializing in environmental chemistry and toxicology at the Virginia Institute of Marine Science and regularly teaches at the College of William & Mary. He also authored the 2017 report on the status of Kepone in the James River. "I do a case study on Kepone in my classes," he said in an interview in 2020, during the Covid-19 crisis. I asked him whether the task force had gotten the ban right.

> I think, given the time, they did a pretty good job. Because I'll ask you the same question—has the Health Department overreacted to COVID? It's the same

question. So it depends on your level of comfort. So, based on the science at the time, it seems like they did pretty good. I think they closed down the river, they waited till the concentrations got below the action level of concern at the time, and then they slowly reduced restrictions based on that. They did the best they could with the data they had. I don't think they overreacted or were negligent in reacting. People want science to be black and white. It's not. You do the best you can with the data you have, and then, as new data comes in, you essentially have to revise your recommendations and your decisions, that's the nature of science.[100]

Funded by the VEE as part of its fortieth anniversary, Unger's 2017 study sampled two important food fish in the James, white perch and striped bass. Researchers found that "65% of the fish analyzed still [had] reportable concentrations of Kepone more than 40 years after the event was first discovered[,] which indicates the persistence of the chemical and how difficult it is to rid a system of a persistent toxic chemical." The good news was that the levels had been "continuing to decline exponentially since 1980" and ought to continue to do so. Funding for toxic monitoring has diminished over the years, though, and it is "more focused on specific contaminants already regulated" than on analyzing unidentified and emerging ones. And while a Kepone advisory remains, one for PCBs in the James is more restrictive, indicating that the issue of toxics in the environment continues.[101]

Residual Governance from Virginia to the French Caribbean

The residues of Kepone left their mark not only on workers and the river but also on the continued regulation of the pesticide and other toxic substances. Once a relatively unknown compound, it came to play a significant role in the development of the environmental management state. At the federal level, along with TSCA, the experience with Kepone helped earn support for the 1977 amendments to the Clean Water Act, including provisions in the legislation requiring pretreatment of toxic pollutants, use of best practices to control runoff and spills, and having the EPA establish an emergency relief fund for toxic spills.[102] These joined a list of other laws that came into being in the 1970s addressing chemical pollution, workplace safety, and hazardous waste. All of them were in part designed to prevent another incident like the Hopewell Kepone contamination from happening. In Virginia, Kepone led to a series of new regulations, including the Toxic Substances Information Act. Moreover, as of 2020, Kepone remains in the river, prompting continued monitoring and study that look to extend into the future.

These developments also resulted in controversy, especially over the ban on fishing in the James River. Kepone showed some of the ambiguity "in the relation between science and environmental values."[103] It also illustrated tensions in applying the precautionary principle to toxic substances. At one level, those involved in regulating Kepone and implementing the ban illustrated the role of science and expertise in a critique of unfettered industrialization and a purely instrumental view of nature. They sought to manage risks and protect human health based on the gathering of data in a dynamic situation relative to the knowledge about Kepone and the economic and political context of the 1970s and early 1980s. These experts and policy makers understood that "uncertainty becomes the ground for additional caution, not business as usual," in the words of *Precautionary Politics* coauthors Whiteside and Gottlieb.[104] Others leaned closer to the instrumental view of science, with pesticides like Kepone part of a "dream of modern development transforming traditional societies," one of countless improvements in modern agriculture in the United States and Europe.[105] In the case of Kepone contamination in Virginia, these experts preferred greater tolerance for pesticides and saw far greater risk in not using them, especially when considering the economic impact on industry and those who relied on the water for their livelihoods. The region's watermen also resisted the increased regulatory burdens and restrictions associated with the environmental management state. For them, expert, scientific knowledge was no match for their experiential knowledge about marine life and the risks of substances like Kepone. Already predisposed to mistrust and skepticism, watermen made common cause with the seafood industry and supportive scientists and politicians in fighting the ban on fishing and harvesting.

The controversy over Kepone and the struggles over the management of toxic substances wasn't limited to Virginia. As important as the focus on Hopewell and Virginia is, as a global substance Kepone connected manufacturing centers like Hopewell with places where growers used it. The pesticide played a role in creating the environmental management state not only in Virginia and the United States but also, as it turns out, in the Caribbean islands of Guadeloupe and Martinique. Even while the United States banned its use, Kepone became a pesticide of choice for banana growers in the islands, and the health effects of decades of use there have been devastating. Knowing what we do now about Kepone in the French Caribbean, regulators in Virginia and elsewhere were right to have been concerned.

CHAPTER 7

The Present and Future of Kepone

The half-life of Kepone takes us into the present and shows us the future. Along with time, Kepone leads us across space, from Hopewell to the islands of Guadeloupe and Martinique and elsewhere. Moving outward from Virginia, the only other documented experience with chlordecone comes from these islands in the French West Indies. Examining their experience adds depth to the Kepone story, traces further the residues of the pesticide, and demonstrates how the chemical relations of Kepone—both environmental and social—connect the events and people in Hopewell with those in the Caribbean. The island tale is more tragic than the one in Hopewell. The poisoning represents the "slow violence" so common in chemical contaminations, incremental damage occurring over a long period of time, decades or centuries.[1] After the United States banned Kepone, it remained the pesticide of choice for French authorities and corporate leaders in the Caribbean, and banana growers used it extensively from the 1970s through the 1990s. As a result, as of 2022, chlordecone poisoning in a clearly chronic, not acute, form, is ravaging both islands, with high rates of cancer and bans on eating food grown in the soil or caught in the ocean. Hopewell and the James River must contend with Kepone's legacy too. There are tougher regulations on polluters, and, with a new sewage system and cleaner waste management practices, the James and other waterways in Hopewell are certainly cleaner than they were. Yet the residues of Kepone linger, as state and federal regulators continue to monitor buried Kepone waste in Hopewell. Moreover, toxic releases and public health remain concerns for the city that strives to rebuild itself as more environmentally conscious.

It's Bananas:
Choosing Chlordecone in the Caribbean

Chlordecone contaminates some sixteen thousand acres of agricultural land in Guadeloupe and nearly thirty-six thousand acres in Martinique. Almost all Guadeloupeans (95 percent) and Martinicans (92 percent) are contaminated with chlordecone. The islands possess the highest rates of prostate cancer in the world, and chlordecone continues to impact pregnant women and their children. In a 2018 visit to Martinique, French president Emmanuel Macron called the environmental contamination of the French West Indies a scandal. "The fruit," he said, "of collective blindness."[2] Yet he refused to discuss demands for compensation for the contaminated population. In the eyes of many locals, it is a colonial crime, the fruit of exploitation of the islands and their populations that began with Native peoples and continued through slavery on the sugar plantations.[3] After decades of agitation from the islands, scientific and medical reports, and various governmental inquiries, in June 2019 the French National Assembly created a commission of inquiry on the impact of chlordecone in the islands, led by Martinican Serge Letchimy. The commission's damning report showed that chlordecone contamination in the islands was not only a "state scandal" but also one that includes the actions of the manufacturers, distributors, and growers, who were "ready to do anything to defend the use of a miracle product without questioning its impact on the environment and health." The report demanded "immediate compensation for all proven damages," including "free care for victims most exposed," a large-scale decontamination plan for the land, and planning to promote an agro-ecological transition for farmers and locals, with an eye toward greater local food autonomy. The report added that fishers restricted from using contaminated waters need compensation and plans for investment. The report, Letchimy noted, was only a first step toward a post-chlordecone world for the islands.[4]

How did it come to this? As political scientist Malcolm Ferdinand testified in the French government hearings over chlordecone in 2019, using chlordecone was a "technical choice" that favored financial interests. Alternatives were available that emphasized agroecological controls without heavy pesticide use that violated "the health of West Indians." While growers claimed that "stopping the use of chlordecone would mean the death of the banana industry in Guadeloupe and Martinique," that has not been the case. The industry moved to more ecological methods once the ban took effect in the 1990s.[5] The technical choice of pesticides began with the intensification of commercial banana growing between 1960 and 1980 as the islands enjoyed a protected market for the French mainland. Control of this cash crop was in the hands of planters with larger landholdings, many of whom were "*békés*," a white minority descended from slaveholders. Both islands still depend heavily on the crop: bananas represent 42 percent of total agricultural production in Martinique and 27 percent in Guadeloupe. The industry accounts for a total of around ten thousand jobs on both islands. In the 1970s, after the French government banned HCH, Dieldrin, and Aldrin, growers relied on Allied's Kepone, which became the major pesticide on both islands from 1972 through 1978, as it proved effective against the weevils in the form of a white powder applied in the soil around banana trees. The intensification of banana growing

A worker spreads chlordecone powder at the base of a banana tree in the French Caribbean. From the documentary *Pour quelques bananas de plus*

sparked a major banana workers' strike in 1974. Along with higher wages, workers demanded the end of chlordecone use and access to protective equipment as they experienced its toxicity in their bodies. The strike ended in violence, with police injured and two protestors killed. Growers refused their demands to end chlordecone and supply safety equipment.[6]

Yet growers were about to face several other problems with their chosen pesticide. Allied stopped Kepone production after the Life Science debacle, and the United States banned its use domestically. News of these events came to the islands as did the 1979 decision by the International Agency for Research on Cancer to classify chlordecone as a possible carcinogen. This sparked France's INRA (National Institute for Agronomic Research) to sponsor research studies in Guadeloupe in 1977 and 1980. These showed chlordecone in the soils and in birds, mammals, and marine life.[7] With Kepone production ending, growers used up all remaining stocks by about 1980. Faced with news of contamination and lack of supply, it may have been a moment to abandon the pesticide. But two hurricanes ravaged the banana crop on both islands, David in 1979 and Allen in 1980. As they operated their plantations again in the wake of both storms, growers faced an explosion in the weevil population. Rather than look for ecological alternatives or even other pesticides, they used intensive lobbying to resume chlordecone use.

Despite the earlier findings regarding chlordecone's health effects, the French government initially avoided the precautionary principle with the pesticide. In 1982 the Martinican company Vincent de Lagarrigue again received approval from the French government to use a new chlordecone formulation, Curlone. They found the supply by a circuitous route. The French company Calliope sold the pesticide, acquiring the patent from SEPPIC, a French company and subsidiary of DuPont. To produce Curlone, Calliope outsourced production to a Brazilian company, AgroCeres, a major agricultural firm, "which itself subcontracted its synthesis through the company AgroKimicos, located in the state of Sao Paulo."[8] This method was common—shifting manufacturing of banned toxics to developing nations as regulations and labor costs increased in industrialized ones. Through this complicated system, the future for chlordecone seemed assured for growers. However, following new European Union guidelines for pesticides, the French government banned the use of chlordecone in mainland France in 1990. Nevertheless, under intense pressure from Lagarrigue executives and planters, the government allowed its use in the islands until 1993. Locals saw this as another example of discrimination against the islands over the mainland. The industry groups argued that they had enough stocks in place for three more years, which would give them time to find an effective replacement for chlordecone. Even so, during this three-year period, growers tried for an extension beyond this period.[9]

During the decades-long battle against the banana weevil, growers spread some three hundred tons of chlordecone powder over some 25 percent of the land surface used for agriculture in each island.[10] The combination of chlordecone's properties and the types of soil in the islands means a long future of contamination. The islands' volcanic soils are divided into three main types: andosols, ferralsols, and nitisols. Andosols trap chlordecone more than the others do, meaning the poison will stay in those lands for much longer. A study done by the French groups INRA and CIRAD (Center for International Cooperation in Agricultural Research for Development) shows that on plantations with regular use of chlordecone from 1972 to 1993, "the molecule will remain in the soil for 60 to 100 years in nitisols, 3 to 4 centuries in ferralsols, and 5 to 7 centuries in andosols."[11] While the Kepone buried in the sediment of the James River might last that long, its presence in fish has declined. Sadly, with its significant half-life and widespread soil contamination, the residues of chlordecone in the French West Indies will continue to be a critical health hazard for hundreds of years at least.

Rediscovering Chlordecone

The 2019 National Assembly report asserts, "Considering the various successive French studies and the Hopewell accident in 1976, which received wide media coverage across the Atlantic, the public authorities should have exercised a duty of vigilance regarding the massive use of organochlorines such as chlordecone."[12] They did not. The situation changed, however, in the 1990s and forced a reckoning with the poison. A major shift occurred when a series of legal and regulatory changes came to France and, by extension, the islands. New laws fixed acceptable limit values for pesticides in water quality analysis and demanded greater attention to drinking water and urban wastewater. Further, the "Barnier law" of 1995 codified the precautionary principle into public policy, allowing for more effective measures at preventing risk from hazards even in the absence of greater scientific proof.[13] Still, this showed that for the "better part of four decades" the French government ignored agricultural workers. Only when authorities recognized that contamination had spread to water systems in the islands, affecting other inhabitants, did chlordecone become a public issue.[14]

Yet even as the French government grew more interested in monitoring water, a regulatory problem occurred that was similar to one that came up in Hopewell. In Virginia, state and federal authorities at first did not account for Kepone since it was not on their list of pollutants. The same was true in the islands, as chlordecone was not on the list of molecules included for detection and analysis. A researcher for the DDASS (Department Directorate of Sanitary and Social Affairs) in Martinique noted in 1991 that he had detected chlordecone in the water samples, but the health authorities did not report this. The reason is that when mainland France banned the chemical in 1990, health authorities in the island removed it from official lists of toxicants—as if removal from a list meant removal from the environment. Only after DDASS released its first report of pesticides in drinking water in Martinique did the agency then include chlordecone. Yet the testing lab, the Institut Pasteur de Lille, lacked the ability to test for it. This is similar to the "slippery" nature of Kepone residues in Virginia in the 1970s, as regulatory systems failed to account for it.[15] There investigators faced early testing difficulties with Kepone in the 1970s, as labs had to develop methods to detect a substance with which they had no experience. Finally, during 1999, island samples went to a different lab in Drôme that could detect chlordecone. They found other poisons as well: the lab reported significant "quantities of organochlorine insecticides residues (dieldrin, HCH beta, and chlordecone), often exceeding the standards authorized," on four sources of drinking water.[16] A study by the Department of Health and Social Development (DSDS) of Guadeloupe, carried out from September 1999 to February 2000, revealed significant pollution of sources in the south of Basse-Terre by organochlorine pesticides banned for several years. Tests detected three compounds at doses a hundred times higher than the norm: chlordecone, HCH beta, and Dieldrin, respectively banned in 1993, 1987 and 1972.[17] All of these were residues of the chemical battle against banana pests.

In 2004 the French Committee on Economic Affairs, Environment and Territory created an "information mission relating to chlordecone and other pesticides in Martinican and Guadeloupe agriculture." This mission resulted in a 2005 report on the extent of chlordecone's use in the islands.[18] In 2007 a damning expose by Martinican intellectuals Raphaël Confiant and Louis Boutrin, *Chronique d'un empoisonnement annoncé* (Chronicle of a Poisoning Foretold) highlighted the extent of the contamination of soils and waterways and the high rates of health problems, including prostate cancer.[19] As Vanessa Agard Jones notes: "As in other instances when their overseas territories have been treated as exceptions to a broader French rule, French citizens of the mainland were deliberately protected from this class of chemicals but their counterparts in *l'outre mer* [overseas departments] were not."[20]

These reports added to growing calls for more action from the French government, which led to the formulation of the first chlordecone plan, in 2008, to coordinate the response among multiple ministries and research organizations. Two other phases followed: 2011–13 and 2014–2020. Many of the environmental and medical studies underway into the 2020s in the islands fall under these plans. Besides research, the plans emphasized improved communications between the various levels of government and local populations, reductions in exposure, monitoring of contaminated areas, cancer registries, and strategies to improve and alter sustainable growing. For going forward, the 2019 report emphasizes more involvement of local constituencies, creating a "space for dialogue, exchange and participation with the populations" in the planning, and more funding.[21] The Chlordecone IV Plan was launched in early 2021, intended to run for six years. With a budget of $92 million euros, the plan covered six overall strategies: communication, research, training and education, health (through environment and food), occupational health, and socioeconomic support for affected professionals.[22]

Magalie Lesueur Jannoyer and others published a series of collected research papers that summarized major findings on chlordecone, allowing for additional comparisons to Kepone in Virginia. With chlordecone in the islands' soil, it leaches into drinking water tables; surface runoff brings contaminated soil into rivers, then to the sheltered bays and the sea.[23] On Martinique, for example, two-thirds of the seventy main rivers showed chlordecone pollution above the regulated standards for water set by the French government (.1 ug/L).[24] About one-third showed sediment contamination, in some cases ranging from less than 10 ppb to 44 ppb. This was less than the James River, which received Kepone waste directly from Allied and Life Science production facilities over a few years; average estimates from 1990 along the lower James were 110 ppb.[25] Testing in the islands also revealed contamination of several marine freshwater species, including goby, freshwater shrimp, and eels.[26] Reports from Guadeloupe showed pollution in all food web components, with concentrations in the freshwater species even higher than those measured in the James River. For example, mullet showed concentrations up to .659 ppm and tarpon up to .499 ppm—both well above the French government level of .2 ppm and those established for fish in the James.[27] In addition, like the James, the biology and feeding pattern of species in the French West Indies influences the amount of chlordecone found in their bodies, with sediment dwellers and those higher on the trophic level seeing the most. Similarly, the general pattern of contamination decreases as one moves from the islands' rivers and bays toward open sea. The contaminated island soil also brings chlordecone to root vegetables and other foodstuffs, especially dasheens (>200 ug/kg or 200 ppb), yams (up to 130 ug/kg), and sweet potatoes up to 100 ug/kg)—all grown for local consumption in backyard systems across the islands. Moreover, animals raised for food in these systems, such as swine, goats, and chickens, also exhibit chlordecone contamination as they ingest vegetables as well as the peel of fruits and vegetables from kitchen waste, and contaminated soil as they feed.[28]

While these contaminated foods are only a small fraction of total consumption in the islands, they explain the chronic chlordecone poisoning of the population. As Phillippe Kadhel and others have documented, the French government has implemented measures to restrict commercialization of these poisoned products. But these measures "only apply to licensed farms and fishermen," Kadhel and co-authors note. "Indeed, there is an extensive informal network of production, distribution, and sale of foodstuffs, as well as families and individuals who produce food for their own consumption. These chains of supply are not adequately controlled."[29] The cruel irony is that the poison used to control banana weevils does not affect the banana fruit, bound for export to the mainland. Meanwhile, chlordecone poisons the food locals use for subsistence.

These research studies led to further connections between France and Virginia. A group of French researchers found that chlordecone displays an unexpected mutability, lowering the time it persists in the environment. However, they learned by testing soil samples from Martinique that the pesticide breaks down into a range of molecules including 10-monohydrochlordecone; 2,4,5,6,7-pentachloro-1H indene; and tetrachloroindene-7-carboxylic acid. Their results show that "chlordecone pollution extends beyond the parent chlordecone molecule" and includes previously undetected molecules, which "illustrates the complexity of chlordecone degradation in the environment" and "raises the possibility of extensive worldwide pollution of soil and aquatic ecosystems."[30] These findings brought some of the French researchers to Virginia, with an agenda to test Kepone in the James River for any similar findings. According to Mike Unger of VIMS, it might be that these breakdown products will remain dissolved in the waters of the James and not accumulate in the food chain since they are "more polar and hydrophilic" than chlordecone, which is hydrophobic and tends to move into animal tissues. This is different from the situation with the soils of Martinique and Guadeloupe.[31]

There is another worrisome issue beyond these areas. The research on Kepone and chlordecone in Virginia and the French West Indies uncovered the fact that the locations of most of the chlordecone produced for export in the world remain unknown. The United Nations Environmental Programme (UNEP) estimates that

Allied and its contractors produced some 1,600 tons of chlordecone from 1958 to 1976 in the United States. Brazil manufactured about 200 tons from 1981 to 1991. Of these 1,800 tons, the French West Indies received about 300 tons, split between 120 from the United States and 180 from Brazil. Where is the remaining 1,500 tons? Relatively small amounts went to Latin America, where the banana weevil was not as threatening as a pest. Some went to Cameroon and Ivory Coast, but how much is unknown. It seems that a significant amount went to Europe. Spiess & Sohn used it to make Kelevan against the potato beetle, shipping the poison inside both West and East Germany as well as Poland and the USSR, including Ukraine. In 1980 West Germany banned its use; East Germany did so in 1983. What issues the residues of chlordecone created in all these other places await study.[32]

Protesting Chlordecone

The loss of livelihood for local island farmers and anglers resulted in a mixture of anger, resentment, and alienation. While the two groups did not communicate directly, their comments echo those from Virginia. One Caribbean farmer, a Mr. Pierce, commented, "In 2006, when I found out my land was contaminated, I had to stop all the basic crops such as yams and sweet potatoes." He went on, "When you listen to the researchers, they tell you that the more often you ingest little doses, the more you poison yourself." "In good consciousness, I could not poison people," he said. "I had to go look for work elsewhere, I worked as a security agent. You become poorer and poorer, your aspirations become smaller. Me, I had the desire to live! Now I don't live, I survive."[33] Striking the same note as Life Science workers, agricultural worker Ambroise Bertin told BBC News: "They never told us it was dangerous. So people were working, because they wanted the money. We didn't have any instructions about what was, and wasn't, good. That's why a lot of people are poisoned." Few were told to wear protective gear, he said. Martinican historian Valy Edmond-Mariette relayed additional fears. "My friends and I were asking ourselves: do we really want children? Because if we give them breast milk, maybe they will have chlordecone in their blood. And I think nobody should be asking themselves this kind of question, because it's awful." Both she and Bertin had cancer: she in her blood and he of his prostate.[34]

The 1974 banana workers' strike was one indication that local workers and health, consumer, labor, and environmental groups were ready to demand more action. Like the workers in Hopewell who used the legal system, those in the islands engaged in a form of biocitizenship, albeit with more social activism, especially after the 1990s. A series of events and reports led to grassroots actions on chlordecone that merged with larger economic, social, and political struggles between the islands and mainland France. Although chlordecone had been banned in 1993, local reports surfaced of growers still using it. French authorities failed to act. Then, in 2002, authorities in the islands seized several tons of chlordecone, almost a decade after the official ban. Clearly, French authorities had acted swiftly to ban its use on the mainland but had failed to do so on the islands. Finally, research by French and island authorities confirmed chlordecone in the environment, and the first medical studies commenced in 2004 to study the presence and effects of chlordecone in the population. Then, in 2006, agricultural workers (who by then had been ignored by the state for a generation) and environmental, consumer, and farm groups in Guadeloupe filed a lawsuit against the French government for failure to protect the public health. In 2007 two environmental groups in Martinique joined them, and these two cases were combined as the issue became one of what Malcolm Ferdinand calls "decolonial environmental justice."[35]

Activism continued. In 2009 a massive general strike spread across the islands and other parts of the French overseas departments. The complaints largely centered on the high cost of food (largely imported) and high rates of unemployment. Part of the focus included environmental issues and public health. Among the demands were reparations for victims of chlordecone. Further protests followed in 2012 and 2013 when fishermen in both countries blocked roads and ports to demand justice for their loss of fishing.[36] Not unlike their counterparts in Virginia, these men complained that chlordecone had disrupted not only their income but also their way of life. One of the leaders, Bertrand Cambusy, demanded "an assistance plan to alleviate the difficulties suffered by seafaring professionals."[37]

More protests erupted in 2018, which prompted a visit to Martinique by Prime Minister Macron. These demonstrations stemmed from a decision by ANSES, the French agency for food and environmental and occupational health and safety, to allow an increase in the limits of chlordecone in meat. As in the United States, science went on trial as part of the political process. Echoing the notion that there is a safe level of toxic substances suitable for human beings, a regulatory change from the European Union over "measurements of liposoluble residues in meat products" resulted in a fivefold-to-tenfold increase in the "safe" level of chlordecone. ANSES found that there was no significant risk associated with the increased levels. Several organizations joined to demonstrate against this, including the collective Zero Chlordecone, Santé Environnement sans Derogations (Health and Environment without Exceptions), and the grassroots group Moun (meaning "people" or "person" in Creole). A prominent leader was the Martinican nationalist Anicia Berton, who connected her own struggles with cancer to the chlordecone pollution of the island. Medical and scientific experts also disagreed with the new decision. Dr. Josiane Jos Pelage of the Martinican group AMSES (Medical Association for the Protection of the Environment and Health) argued that, since chlordecone is an endocrine disruptor, "its mere presence is sufficient to possibly

Demonstrations in France against chlordecone. The public health crisis from chlordecone in Guadeloupe and Martinique prompted large demonstrations against the French government both in the islands and in Paris. Here demonstrators demand justice at a protest in Paris in February 2021, taken when mask restrictions were in place due to the coronavirus pandemic. Photo by Karim Ait Adjedjou/Avenir Pictures/Abaca/Sipa USA (Sipa via AP images)

create major harm." Another group, Ecology Urbaine, led by Louis Boutrin and Raphael Confiant, noted that chlordecone is just one of several toxic substances used on banana plantations, so deriving a limit for one does not take into account the combined effects of multiple exposures to multiple substances.[38]

As all of these developments went forward, the 2006–7 combined lawsuit languished in the court. Then, in January 2021, French judges suggested that the statute of limitations on the case had run out and that some files for the case had even disappeared in the intervening years. This contrasts with the Kepone story in Virginia, where both Allied and Life Science faced immediate hearings and a series of legal judgments. An angry response from island residents followed, with some support among mainland political leaders. Pascal Tourbillon of ASSAUPAMAR (Association for the Preservation of Martinican Heritage) expressed his dismay: "It has been dragging on in court for 14 years and today we are told that the facts are time-barred? It is not our fault!"[39] Local political leader Francis Carole, head of PALIMA (Party for the Liberation of Martinique) stated: "Thousands are mobilized to respond to that gob of spit the French state is sending us, the statute of limitations."[40] French MP Olivier Serva, representing Guadeloupe, started a petition against dismissal. Dismissal, the petition argued, "would be a humiliation for Martinicans and Guadeloupeans." "Let us fully subscribe to the victims' fight for the legal recognition of their suffering and for their reparation in Guadeloupe and Martinique."[41] The organization Lyannaj pou depolyé Matinik started an awareness campaign of the problem of chlordecone in 2020. The collective, which brings together various local associations and trade unions, has organized demonstrations in Martinique with the slogan "Refuse the denial of justice."[42] They have also included a list of fifty-two demands, some that would include reparations for victims of chlordecone.[43] By bringing together various groups, the demonstrations connected social and political issues associated with colonialism to ecological ones. Some have entered party politics. In 1991 the Greens Guadeloupe formed. In 1992 former members of ASSAUPAMAR founded the political party MODEMAS (Movement of Democrats and Ecologists for a Sovereign Martinique). Another party formed in 2011, Martinique Ècologie, founded by Louis Boutrin and Raphael Confiant. Between activism and parties, the issue of chlordecone manifested into these broader political movements and remains central to both islands. They also show similarities and differences in how biocitizenship operated in Virginia and the Caribbean.

Human Effects of Chlordecone Poisoning in the Islands

At the core of the activism are the real and potential health effects from chronic chlordecone exposure. In this tragic way, island residents and Life Science workers are connected, bodies at risk at different points in the Kepone chain, each bearing the chemical burden of the pesticide. There are similarities in the health concerns but also important differences. Hopewell's white, working-class men suffered from a brief, high-level poisoning. With time and interdiction by medical professionals, their bodies could remove most of the Kepone. Although it is uncertain, it appears most did not suffer long-term health problems from Kepone. In contrast, Kepone uptake by islanders of African descent invokes the direct legacy of slavery and colonialism. The French state allowed chlordecone on the islands, *l'outre-mer*, while restricting its use in France, the metropole. In Guadeloupe and Martinique, men, women, and children ingested chlordecone over years, and even newborn infants were affected, adding to the cycle of slow violence in the islands. Indeed, to borrow the ideas of Norah MacKendrick and Kate Cairns, chlordecone poisoning in the islands illustrates in a stark way how everyone in industrial society carries a chemical body burden, with newborns coming "pre-polluted" into the world. The question of pregnancy and children, as Valy in Martinique noted, brings forward powerful ideas associated with gender, reproduction, and maternal responsibility. For men in Hopewell, they expressed the fear of sterility. For them, what was feared did not happen, but unfortunately this has not been the case for women in Guadeloupe and Martinique.[44]

The hospitalization of male Life Science workers generated an initial focus on Kepone's effects on male reproduction. Since then, the scientific community expanded the data in the coming years to include animal studies on females. From these, it became clear that chlordecone interfered with normal fetal development. Chlordecone crosses the placental barrier and may be transferred to newborns through breast milk. Chlordecone is also an endocrine disruptor, leading to the likes of decrease in fertility rates, problems in embryo development, reproductive abnormalities, and cancer.[45] More direct knowledge on the effects of chlordecone on women in the French West Indies came from the Timoun Mother-Child Cohort Study (2004–7) on pregnant women and newborns in Guadeloupe. Data from this study formed the basis of further research that showed prenatal exposure to chlordecone is associated with shorter gestation and increased risk of premature birth and lower birth weight.[46] In early childhood development, newborn Guadeloupeans show a reduction in fine motor skills and short-term memory—consistent with the effects on Life Science workers. Yet the effects in utero and among newborns occurred at levels roughly one thousand times lower than that of the exposed Hopewell workers, indicating that the prenatal period is one of higher sensitivity to the endocrine effects of chlordecone.[47]

For men in the islands, additional research is underway examining the connection between chlordecone exposure and prostate cancer. According to the World Cancer Research Fund, as of 2018 Guadeloupe and Martinique are first and second in the world for age-standardized rates of prostate cancer.[48] The first major investigations among men on Guadeloupe, the Karuprostate study, showed a significant increase in the risk of prostate cancer with higher rates of chlordecone in the blood.[49] Others, though, found no direct link between incidences of prostate cancer and chlordecone, so there remains a need for additional studies.[50]

The story of chlordecone will continue for island residents as they battle for environmental justice. These residues show how the "past has been built into our chemical environments" in the present. Following the chemical relations of chlordecone led from Hopewell to Europe and then the Caribbean, with stops in Brazil, other parts of Latin America, and Africa. The island residents of Guadeloupe and Martinique now must address the extensive half-life of chlordecone in the ground, water, food, marine life, and their bodies. This stems from not only its chemical formula or the peculiarities of soils or animal species but also the decisions associated with the production, distribution, and use of chlordecone and the failures to regulate or "clean up the leftovers."[51] These island residents are not alone. While there are some critical differences, those in Hopewell also addressed the half-life of Kepone as they moved forward from the 1970s. It is to them we now return.

Hopewell after Kepone: The Residues of History

Hopewell had trouble getting out from under the Kepone cloud. As the trials continued against Life Science and Allied, the city moved forward in improving its wastewater plant. Plans were underway to construct a new regional facility, and construction had begun when Kepone hit. As Mark Haley recalled in 2016: "A lot of people always thought that the treatment plant in Hopewell was an aftereffect of Kepone, but actually it wasn't. Because of the complexity of the plant, the design, the notion of going to secondary treatment was a nationwide effort. The Clean Water Act was promulgated in 1972, so a lot of cities were only doing basic low-level treatment and basically using the lakes and the rivers as the final step."[52]

The new plant would take not only the city's waste but also waste from Hopewell industries as well as Fort Lee. Unfortunately, its opening only further marred the city's reputation. The EPA paid for half the cost, with the rest of the money for the $44 million project coming from local industries, Fort Lee, and the State of Virginia. It seems clear that those in charge mismanaged the construction. Pressure from government may have led to the facility opening earlier than it should have. When it did so, in August 1977, it experienced immediate problems with treatment of industrial discharges and equipment failure. Virginia's State Water Control Board then began periodic inspections and directed the city to take action, as did the EPA. Still, the facility exceeded its NPDES permit from October 1977 to February 1979. Both the VEE's Jerry McCarthy and board director Bill Cummings weighed in as well. McCarthy pressured Governor John Dalton, "Let us know what action you intend to take to correct the problem at Hopewell and to prevent the State Water Control Board from dragging its feet on problems such as this in the future, lest we have to learn the lessons of the Kepone disaster once again."[53]

The city's probation in the federal Kepone case required that it refrain from violating federal, state, or local laws. City leaders, including newly hired city manager Clint Strong, believed the Kepone saga led the EPA to single out Hopewell among the many municipal violators of clean water requirements. Strong was Hopewell's city manager from September 1977 to July 2000. "We finally got the General Accounting Office in Washington to do an investigation of EPA," he said.[54] The General Accounting Office (GAO) report came at the behest of Virginia senators Harry F. Byrd Jr. and John W. Warner Jr., and Congressman Robert W. Daniel Jr. According to the report, Strong and others had good reason to believe what they did. The EPA's regional director told the city, "The relationship between the 1976

Kepone incident and the plant's current compliance and funding problems is not merely geographical, as you assert."[55] Further, an EPA attorney assigned to the suit told the *Washington Post* that the EPA investigation was prompted by "the potential importance of the plant. We remember Kepone. It is a sensitive issue."[56] The nationwide review from the GAO "looked at 15 plants determined by EPA to be "worst case situations" in three EPA regions and found that EPA's enforcement action varied from none to minimal, followed no particular pattern, and was not as timely or as effective as it could or should have been."[57] Allowing Hopewell to further violate environmental laws would be another political mark against the EPA, something that perhaps would not have been the case with other communities engaging in similar behavior. There was another, somewhat humorous connection that former city manager Mark Haley recalled. "There were some surveyors that had a sense of humor, because there are some benchmarks down at the treatment plant that are named 'Kepone 1' and 'Kepone 2.' One's right in front of the admin building, right on the sidewalk... one of those little disks that the surveyors use to start when they want to do something. It's labeled as 'Kepone 1.'"[58]

After this inauspicious beginning, Hopewell modified the plant and met the requirements by the spring of 1980, but the EPA presented the city with a consent decree that included paying of fines for past violations in addition to fixing the problems. When the city refused to sign, both the EPA and the State of Virginia filed a lawsuit against the city in August 1980. Eventually the city, EPA, and the State of Virginia signed a new decree in 1981. Hopewell agreed to pay a $25,000 fine and an additional $10,000 toward an environmental project or projects in the city, and to improve pretreatment of waste to avoid additional problems at the facility.[59] The facility changed its name to Hopewell Water Renewal in 2016 and added an innovative, award-winning system to remove nitrogen in continuing efforts to maintain the health of the James River and the Chesapeake Bay. As Strong said: "The process was made to work, and now it's one of the best in the state."[60] Perhaps a sign of progress, Haley noted in 2016, "The eagles are back, and they're happy, because they can fish the James now. There's nested eagles all around here, all over Prince George. We have them at the treatment plant."[61] Indeed, in 2019 there were three hundred pairs of eagles along the James River, which was the recovery goal for the entire Chesapeake Bay.[62]

Besides a new regional plant, Hopewell leaders sought other ways to minimize the chance of another Kepone and rebuild the city's image. Strong remembers the general sense at the time. "How do we get out of this situation where everybody in the state, we thought, was down on us—all the state agencies? If they wanted to make points, they blasted Hopewell, and everybody outside was picking at Hopewell, and that makes people very defensive. So we created in the early eighties an Economic Development Committee composed of twenty-one citizens who were going to look at our image, look at what we're doing to improve ourselves. This group came up with a whole series of recommendations. It was schools. It was streets. They wanted to have it all fixed up. It was controversial." The group published its report in 1984. The city council initially resisted, he recalled, but over time began approving money for projects. "All our schools were redone, streets were all overlaid. The regional plant was working, so we were in pretty good shape."[63]

But tension remained between the city and the industries, part of the Kepone legacy. "The plants," Strong remembered, "would have a spill, but they wouldn't tell us." They would tell "the feds," he said. "The feds wouldn't tell us either, so we finally got upset to the point where they were endangering our people. Nothing happened that severe, but it could have."[64] It is not clear what events occurred that he did not consider severe. The disagreements got heated, and eventually, in early 1991, Strong, other city officials, and the plant managers met at a retreat to hash out differences and come up with a solution. The feeling from Strong (and others): "We were a smokestack community with constant environmental concerns—none of which were being addressed." For their part, industry representatives felt the city had "applied too much muscle in the environmental respect," said Strong. "I admit, I did hit them kind of hard on some things, and I don't apologize for it."[65] After the meeting, Strong recalled, "[The group] decided that we needed to have an ongoing dialogue, and so we set up what was called the Hopewell Community and Industrial Panel." This panel was made up of plant managers, the city manager, and selected citizens to promote cooperation and communication between industry and the city. "Things really got better after that."[66] The plants became more involved in the community through a variety of programs, including plant tours. The managers held a town meeting in 1992 to answer questions from the community and went on to publish a regular column in the newspaper addressing continued questions. The goal was to "show how ecology and economics can combine for jobs."[67] The EPA praised Strong and the city in the 1990s for the turnaround. Editors of the *Hopewell News* proclaimed, "Hopewell is no longer a name to be whispered in conversation dealing with the environment."[68] Similar organizations emerged as well. There is the Hopewell Emergency Network System to enable quick communication across the city in case of an emergency. Hopewell also created a Local Emergency Planning Committee that meets regularly to review procedures and enact drills.

As the islands of the French West Indies are addressing decades of chlordecone poisoning, the city of Hopewell is coming to terms with its own legacy associated with Kepone and its role as an industrial city. While Kepone is no longer as visible or direct a threat in Hopewell as chlordecone is in the Caribbean, the legacy and memory of the events in the 1970s live on. The city, once the "Chemical Capital of the South," is on a path to rebranding itself as eco-friendly. Clint Strong claimed: "There's a positive side to Kepone, and we became more environmentally conscious."[69] Assistance for various projects came from government and other outside organizations, including the Virginia Environmental Endowment.[70] Yet Hopewell remains an industrial place. And Kepone remains in the soil, contained in engineered burial pits that require regular monitoring to ensure that it does not leech into drinking water or add to the sediments already in the James River. Health rankings show that the city struggles to maintain a healthy population and environment. Several recent efforts to improve these areas show citizens and leaders coming to terms with the city's industrial past and present. In Jeanie Langford's view, "Hopewell does strive to see that industry and nature can live side by side."[71] Along with waste treatment and emergency issues, city leaders moved toward revitalizing other aspects of the economy. One focus has been on downtown, centered on renewing the 1928 Beacon Theater and the public library. The City Marina also saw new investments, as have a number of other businesses.[72] One less formal effort to revive the city's reputation among its own residents came from Kit Weigel. Weigel, who passed away in 2020, started her "Happy in Hopewell" Facebook campaign to inspire and encourage positive actions and views of the community. She even had stickers and buttons made. Another major project, spearheaded by the nonprofit FOLAR (Friends of the Lower Appomattox), was the construction of a recreational trail, the Riverwalk along the Appomattox River. It opened in 2019 and received an Environmental Excellence Award from the governor's office in 2020. It is slated for further development.

Of Sediments and Sludge:
Monitoring and Disposing Kepone Residues

As the city moves ahead, residues of Kepone remain in Hopewell and in the sediment of the James River. Along with the Kepone task force, Virginia coordinated with VIMS in creating a Kepone monitoring program for James River marine life that continued into the 1990s. The 2017 report from Mike Unger at VIMS noted that while the fish sampled showed levels of Kepone below the action level, Kepone nevertheless remained in the river decades after the crisis. While the useful life of Kepone was over (at least in Hopewell and the United States), its half--life continued, ensuring that the time, work, and costs associated with its disposal went forward into the future. The process of monitoring Kepone residues in Hopewell involved local, state, and federal government officials, along with those from Allied and its subsequent owners Honeywell and AdvanSix (a Honeywell spinoff). Regarding Kepone in the James River, government officials studied the issue and decided to leave Kepone where it was and allow the river's movement to continue covering the pesticide with sediment. To address the waste left behind in Hopewell, officials examined proposals to incinerate the sludge from the treatment plant, but local resistance and high costs dissuaded them. The solution was to bury the sludge along with remnants of the Life Science facility and Allied's Semiworks plant that once made Kepone. There are three sites on or near the city's landfill and old sewage treatment plant and three on the property of Allied (now AdvanSix). Each site falls under state and federal monitoring to ensure no Kepone seeps into the surrounding waterways.

As the immediate Kepone crisis unfolded, the EPA coordinated studies to determine what to do about the Kepone in the James River. These involved several experts from various agencies, including the Army Corps of Engineers, the Department of Energy, the EPA, and VIMS, as well as government leaders in Virginia and Maryland. The group also exchanged information with New York's PCB taskforce addressing contamination in the Hudson River. In New York, dredging commenced to remove PCBs from the Hudson, but the EPA's 1978 Feasibility Study for Kepone recommended against that approach: "Based on the enormous costs of total James River amelioration efforts, the lack of knowledge on ecological impacts of widespread mitigation efforts, the unavailability of economic impact determinations and supportive evidence that most of the Kepone will remain in the zone of turbidity maximum, no full-scale cleanup action on the James River should be undertaken at this time."[73] Estimates of the total cost ran as high as $7.2 billion. The study also noted that there was "no imminent danger of Kepone contamination" to the Chesapeake Bay. Modeling and field sampling from the river indicated that Kepone would remain largely in the same area of the river, within a "sediment trap," as opposed to spreading more widely as PCBs were likely to do. There had been a fear that flooding from hurricanes like Agnes in 1972 might dislodge the Kepone. In fact, those events increase sedimentation rather than send it downstream and into the Chesapeake Bay. Of course, the trade-off was that Kepone remained in the river and will continue to show a presence in marine life for years to come. Birds of prey and humans eating fish will also ingest Kepone, even if in doses well below the published action levels.

As studies for addressing sediment went forward and problems with the new sewage plant unfolded, more controversy erupted over plans to dispose of what was left of Kepone and its production. Allied's remaining Kepone stocks, some 85,000 pounds, could have been sold overseas, but the company decided to remove it from the market and looked for ways to either bury it or burn it. It remained in black drums at the Race Street facility in Baltimore and at the former sewage plant in Hopewell. Allied also needed to dispose of engineering material from their own manufacturing of Kepone. Then there was the matter of some nine million pounds of Kepone sludge left over from the sewage treatment plant, a matter handled by the EPA and Virginia's Health Department.

In 1977 the EPA funded a test at a facility in Toledo, Ohio, that showed the possibility of burning Kepone and sludge without creating hazardous fumes. Based on this, Virginia's Board of Health gave serious consideration to a plan to burn the Kepone and sludge using a portable incinerator brought to Hopewell. The board held two public hearings on this, one in Richmond and the other in Hopewell, in January 1978. At the Hopewell hearing, an engineer for the company overseeing the process stated that burning was "safer than moving Kepone and risking exposure of it." Yet the Hopewell City Council had already adopted a resolution on December 20, 1977, rejecting incineration, as did the surrounding Prince George County.[74] Despite the resolution and public opposition in Hopewell, the Board of Health approved using the incinerator at its meeting on February 15, 1978. The decision then went to the state's Kepone task force for their recommendation to Governor John Dalton.[75]

While these plans were underway, Allied sought a solution for its own Kepone waste. In November 1976 the State of Idaho rejected a request to bury some five thousand pounds of non-burnable Kepone (laced with arsenic), sixty tons of sludge, and five tons of Kepone materials in a deactivated Titan missile chamber south of Boise already used for other waste.[76] In 1978, as it awaited Dalton's decision, Allied contacted Rechem International to run a test burn of Kepone at a facility in Pontypool, Wales. But townspeople there and local environmental groups "threatened to besiege the local plant of Rechem International," licensed by the British government to dispose of toxic waste. A Wales MP declared: "South Wales is not prepared to become a receptacle for the excreta of an incontinent sector of American capitalism."[77] The government banned the importation of the Kepone.[78]

Apparently Governor Dalton was waiting to see the results of this test in Wales before making a final decision on approval to burn Kepone in Hopewell. "Before any recommended method of mitigation is accepted, it will be essential that I know there will be no harmful effects on our people or our environment."[79] The Kepone task force had also not made a final recommendation by the time Britain banned the test. Only a few days later, on April 7, Virginia's Kepone program coordinator Roy Puckett stated to the press that the state had "in effect" decided not to burn Kepone in Hopewell after local opposition, even though Governor Dalton had not issued a formal statement. Locals were happy. Then Mayor Hilda Traina said she was delighted with the news but still wanted Kepone out of the city. Reverend Curtis Harris had opposed burning it, arguing that the incinerator might have an undue effect on the African American population in the city that already lived closer to the landfill and the chemical plants.[80]

Dalton delayed a formal decision into the summer of 1978, and the point became moot. Sometime in July, Allied entered into an agreement with the German firm Kali and Salz to ship its remaining barrels of Kepone, sitting in its Race Street facility in Baltimore, to Germany for burial in Herfa-Neurode, a former salt mine in Hessen that is the world's largest underground landfill for toxic waste. Allied and Maryland state officials traveled to Germany to inspect the facility and meet with German officials. The first shipment of 504 black drums of Kepone set sail from Baltimore aboard the *Düsseldorf Express* on an August afternoon in 1978.[81] But the transfer met with some resistance in West Germany. Allied suspended the shipments after a few weeks as coverage in the West German press raised concerns, and the group *Pro-Gruen* (For-Green) that would soon be part of the Green Party sent a message to the chancellor urging reconsideration and full inspection by the state government in Hesse.[82]

Negotiations over continuing shipments lasted through early fall. On September 22, Allied officials, along with those from Hopewell and state government, met and planned a move for the remaining five hundred barrels of Kepone sitting at the former Hopewell sewage plant. Presumably, the plan was to ship them to Germany also once the suspension was lifted. Then, with no publicity, during the darkness on Friday, December 1, 1978, the remaining barrels in Hopewell were loaded onto seven tractor trailers and sent to the Virginia Port Authority Terminal in Portsmouth, escorted by two state police cars and an airplane. There they were loaded onto a ship bound for Germany. An Allied spokesperson stated that this completed "the company's previously announced agreement with the West German firm" but was "less than the amount previously authorized by the German government." It is not clear if this meant that Kepone stored in Baltimore still remained there or not. Allied had notified Virginia's secretary of commerce, Maurice Rowe, on late Thursday afternoon, the day before, and Rowe had then arranged the escort. The city manager in Hopewell knew Friday when the loading began. The state's Kepone coordinator, Roy Puckett, only learned of the move late Friday night when he received a call from Allied that it had been completed. Puckett wasn't concerned. "I'm just happy the stuff is out of Virginia," he said.[83] Kepone left Hopewell the way it came in, with secrecy and little public knowledge.

With Kepone shipped out of the country, the work continued on what to do with the waste left in Hopewell. Allied's settlement with the State of Virginia covered the initial costs for these residues. As noted, there are three Kepone sites on or near the city's landfill and old sewage treatment plant. One site, labeled SWP 192 (Solid Waste Permit 192) under Virginia's Department of Environmental Quality, is a closed sanitary landfill, about one acre in size, surrounded by a chain-link fence topped with barbed wire. It is located on land of the former Hopewell landfill. After getting a permit from the Department of Health, on June 30, 1976, the site received about one thousand cubic yards of Kepone-contaminated "soil, concrete pavement, gravel, clothing, shoes, gloves and filters" from the Life Science cleanup operation.[84] Available records from Virginia's DEQ show that the State Water Control Board installed a monitoring well, though it is not clear how long the agency recorded data or what they found. A second area, SWP201, is located on the older southeast section of the landfill. This encompasses the hillside where Life Science workers dumped Kepone waste toward Cattail and Bailey Creek. Monitoring there as part of the actions against Life Science discovered that leachate from the site contained Kepone that entered into the waterways.

The landfill had a negative history associated with it even before Kepone. On August 6, 1966, as Martin Luther King Jr. led marchers through Chicago, Hopewell's black residents—led by Reverend Curtis Harris—protested the proposed location for the landfill adjacent to Rosedale, a mostly African American neighborhood in the city. A group of about 40 protestors marched from Rosedale to the city hall, where they were met by some two hundred members of the Ku Klux Klan. The Hopewell police force—all twenty-two members—kept a lane open as the protestors walked between lines of Klan members (men and women) to present their demands to the city. On the steps of the Hopewell Municipal Building, the protestors sang, prayed, and shouted "Freedom now!" as the Klan retorted "Never, never!" and "Nigger, go north!" while a crowd looked on. After some tense moments, Harris led the protestors back to his church. Despite the protest, Hopewell opened the landfill a month later.[85] Once it was operational, the landfill had problems with drainage flowing onto Rosedale streets.[86] Kepone added to the problems. To deal with the leachate, in the early 1980s Hopewell regraded the hillside and installed equipment to divert the leachate from the groundwater.[87] A third site, SWP 271, is also behind a fence topped with barbed wire. A burial pit sits inside a three-acre site on land at the original sewage treatment plant that contains some forty-two hundred cubic yards of the Kepone-laden sludge and other materials from the old treatment facility, dumped there in 1979. In 1983 the Health Department placed the site in an inactive status, likely because sampling results to that point showed little or no Kepone in monitoring wells.[88]

Passage of two federal laws prompted additional identification, remediation, and monitoring of waste on Allied's property. The Resource Conservation and Recovery Act (RCRA) of 1976 governs the disposal of hazardous waste, generally in ongoing waste streams. The Superfund legislation in 1980 added further scrutiny in cleaning up hazardous waste remaining on the property.[89] Allied reported

Kepone sign in Hopewell, Virginia. The residues of Kepone are in many areas, including this hazardous waste disposal site that contains Kepone-laden sludge from the now-defunct sewage treatment facility. Other waste sites contain the remnants of the Life Science facility, the waste from that plant, and the waste and the physical remains of the facility where Allied made Kepone before contracting out with Life Science. Photo by Gregory Wilson

hazardous waste on its property, which included areas that contained Kepone residues and remnants. Under the Corrective Action Program of the RCRA, the EPA identified Solid Waste Management Units (SWMUs) on the Allied site that required remediation and monitoring. Three of the fourteen sites on Allied property related to Kepone.

SWMU 1 is an unlined, diked impoundment of dredge spoil on Allied property where Poythress Run meets the James River. Some 810,000 cubic feet of the dredge spoil from that area contained twenty-nine pounds of Kepone.[90] Allied capped the impoundment and seeded it with grass. Studies done under contract for the EPA indicate that any water that might leach from this site would contain low levels of Kepone, as the pesticide has a low level of solubility in water and high sediment or soil absorption.[91] SWMU 3 is a coal ash pile with remnants of some 350,000 gallons of Kepone spray. Once Life Science closed, the structures were rinsed off to remove Kepone. That water was filtered, tested to ensure Kepone levels at or below 10 ppb, and then stored in railcars. Allied received approval from the Department of Health to spray that water on piles of coal ash on Allied property. Tests of the soil in the pile done by Allied and the Virginia State Laboratory in the 1970s showed Kepone ranging from .017 to .23 ppm.[92]

SWMU 27 is an engineered landfill, three hundred by two hundred feet, and seventeen feet deep, close to Route 10. It contains the remnants of Allied's Semiworks plant that manufactured Kepone and other products. In 1977 the EPA and Virginia's SWCB and Health Department ordered the facility closed, and the consent decree required Allied to construct the landfill, which contains some thirty pounds of Kepone.[93] Allied (and then Honeywell and AdvanSix) monitored the landfill and reported the results to the state. Data from the state DEQ shows that only seven times was there enough liquid present in the monitoring well from the landfill to be sampled. Each time Kepone levels were below the detection limit, which shifted from 10 ppb to 1 ppb. The last attempt at sampling occurred in 2020, and there was no water from which to sample.[94]

These reports on Kepone disposal and monitoring in Hopewell, and those regarding its shipment to and disposal in Europe, take us back to residues. The need to monitor and test reminds us how Kepone and other chemicals are transgressive, disobeying even the best technical solutions designed to contain them. As historian Martin Melosi notes, disposal of Kepone with other waste in landfills or former mines "constitutes a problem never solved in the past and unlikely to be solved in the future."[95] In addition, defining the Kepone in the barrels and Kepone manufacturing equipment as waste meant that their useful life was over, but addressing the waste required considerable expenditure, energy, work, negotiation, and engineering—all of which continues. Much like the pesticide itself, the waste left behind entered the geopolitics of the environment, whether in Virginia, Europe, or the Caribbean, as various individuals, agencies, companies, and organizations cooperated and challenged one another over what to do about the remnants from Kepone production.

The Work Remaining

Environmental monitoring remains an issue more generally for Hopewell since balancing industry and the environment is an ongoing act. Kepone was but one example of a larger history for Hopewell's identity as an industrial community. While the city has made great strides in protecting health and the environment, significant work remains. The city typifies fenceline communities across the nation where residents, largely made up of lower-income whites along with racial minorities, live near polluting industries and risk increased exposure to toxic chemicals.[96] Hopewell sits in the Virginia Health Department's Crater District, named for the famed Battle of the Crater during the siege of Petersburg in the Civil War.

Cancer rates are high there. Incidence rates as of 2019 showed that the Crater District was the worst in the state, with a rate of 494.1 per 100,000 (Virginia's overall rate was 417.9), and ranked thirty-second out of thirty-five districts in mortality rates. Much of this is skewed by race, with African American rates higher than white rates.[97] Hopewell's average rate from 2013 to 2017 was 526.3, among the highest in the state.[98]

Air toxins also remain a serious concern for Hopewell. A 2017 report by the Sierra Club of Virginia that used data from the federal Toxic Release Inventory (TRI) ranked Hopewell first among Virginia municipalities in toxic air emissions, with some 3.4 million pounds released in 2015.[99] The 2018 TRI shows Hopewell third among Virginia jurisdictions for on-site releases of toxics at 2.26 million pounds, or 7.99 percent of the state total. Hopewell's AdvanSix Incorporated was fourth-largest facility in the state in on-site chemical releases, some 1.34 million pounds.[100] AdvanSix, the former Allied facility, has been targeted for several violations of the Clean Air Act, at one point leading to a 2018 raid on the site by the EPA, the Virginia Department of Environmental Quality, the FBI, and the Virginia State Police.[101] Other companies in Hopewell, including Hercules and the paper company Stone Container (which was acquired WestRock in 2011) have also violated clean air laws.[102] Putting the respiratory illness and cancer risk from air toxins together with economic data for Hopewell shows how the city comes under concern for environmental justice. Racially, the city's population is almost evenly split, with 50.9 percent white and 43.5 percent African American. The state's 2019 median household income was $74,222, while Hopewell's was $39,030. Per capita income in Virginia is $39,278, and Hopewell's is $21,927. The poverty rate in Virginia was 9.9 percent, while Hopewell's rose to 23.6%.[103] These startling statistics demonstrate how places like Hopewell continue to absorb the toxic costs of national and global industrial production and consumption.

There is some good news for the James River, at least. It is much cleaner than it was in 1975. As the VEE's Joe Maroon noted, "[For] too many years, the James, even right here in the city of Richmond, was not viewed as an asset, a community asset. Now it is. It's turned around to the point that a few years ago the city of Richmond was named the best outdoors town in the country, because of the James, the James River Park, the trails, all of the different things that have refocused the community back towards the river. The river's an asset now, not an eyesore, not a dumping ground."[104] Judge Merhige's son Mark agreed. "The recent success of Richmond has been based around our river, to a great extent."[105] However, much work remains. The James River Association (JRA) formed in 1976 in response to growing public awareness of the polluted river. In its 2019 report, "State of the James," the river earned an overall grade of 60 (out of 100), a B- in the JRA's grading scale. Signs of improvement are there. Bald Eagles nest along the river, and in 2018 sturgeon returned. Yet seemingly intransigent issues remain, including nutrient overloads of phosphorous and nitrogen from industry and agriculture, stormwater runoff, and sediment pollution. Despite shad hatchery programs, the shad population remains at historic lows as of 2021. Climate change may make it harder to move these indicators in a positive direction, if increased rainfall and flooding become the norm for the area.[106] As of 2021, the "Fish Consumption Advisory" remains in effect. According to a report by Noah Sachs and David Flores, the toxic residues along the river pose another threat. "More than 2,700 industrial facilities regulated by federal and state programs for toxic and hazardous chemicals are located in the most socially vulnerable census tracts in the James River watershed." Of these, Sachs and Flores conclude, "More than 1,000 of these facilities are exposed to potential river flooding, hurricane storm surge, and/or projected sea-level rise." In addition, they say, "[O]ver 473,000 of the 2.9 million people who live in the James River watershed are in communities that the U.S. Centers for Disease Control and Prevention defines as 'socially vulnerable' to disaster and that also contain flood-exposed industrial facilities with toxic substances."[107] Much progress has been shown for the James River and Hopewell since the 1970s, but much more remains.

EPILOGUE

The Kepone crisis was an important episode in recent environmental history whose effects continue to reverberate. The environmental management state evolved in the wake of Kepone's production, distribution, use, and its residues, not only in Virginia but also elsewhere in the United States, Europe, the Caribbean, and beyond. The crisis exposed corporate malfeasance and weaknesses in government regulation. Fortunately, there were those who reported what was happening and acted to contain the contamination and address the poisoning, even as others sought to dismiss the danger. The debates over responsibility and regulation revealed the ways uncertainty operates at the nexus of science and policy making. The legal system brought some restitution and set precedents for later decisions. In the form of the Virginia Environmental Endowment there emerged an organization devoted to addressing environmental protection. In the islands of Guadeloupe and Martinique the crisis developed over a longer period, led to a larger public health emergency, and revealed even greater deceit and betrayal. The outlook there remains grim.

These issues linger in the memories of those interviewed for the book. Several reflected on their importance and meaning. Joe Maroon, director of the Virginia Environmental Endowment went back to his time leading the Chesapeake Bay Foundation. For him, the growing action in the 1980s against pollution of Chesapeake Bay built on Kepone: "The Kepone awareness... I think touched people in ways... they had never before made that connection." Maroon, who grew up in Akron, Ohio, likens the memory and effect of Kepone to his own recollections of the infamous 1969 Cuyahoga River fire in Cleveland. People of his generation who were in Hopewell or Virginia remembered each one. "If I go talk somewhere and I mention Kepone, they immediately have a reaction to it. It stayed with them. It was like the Cuyahoga River for me. It was Kepone for them, and that was a big deal."[1]

When asked about the legacy of Kepone, David Paylor, former head of Virginia's DEQ, reflected on both his own memory and the impact on environmental regulation more generally in Virginia. Like Maroon, Paylor noted the connection to the Cuyahoga River fire. "It's the Cuyahoga River for Virginia, in terms of the public conscience. The workforce is turning over, and so folks who lived it, and there aren't many of them left in the agency, but it lives in infamy, not just within the agency, but also within the thinking of... the environmental community or the public. Folks continue to look back to it as a benchmark, catastrophic benchmark, and an example of that—and so I think probably within the agency, it's fading, but it keeps coming up." For example, he said, "Many of the monitoring systems and so forth that our agency developed over the years probably were initially informed by things that we learned with all of this." Sometimes he will also hear public comparisons of current issues back to Kepone, which suggests that Kepone remains in the collective memory as well. When there are large enforcement decisions, for example, "we'd go out and get public comment on it, and a lot of times the public comment is within the context of Kepone. 'Well, this is almost as bad as the Kepone is, so you ought to be charging them that many dollars in today's dollars'—[that] sort of a thing." Referring to a coal ash spill on the Dan River in 2014 that impacted Virginia's portion of the watershed, Paylor commented, "You hear some folks saying, 'Man, this is the worst thing since Kepone,' so it's there. It's just like the big-deal thing."[2]

Overall, Paylor reported seeing environmental progress since he started working in the field right before the Kepone crisis. He maintained that view despite criticism of the DEQ from environmental groups who have argued that the agency is too easy on polluters and noted that "since 2001 DEQ's general fund appropriations have been reduced by $37 million per year, and 74 positions have been lost."[3] For the first two or three years of his job, Paylor said, he was investigating fish kills from uncontrolled industrial and municipal discharges. "The bay's a lot cleaner than it was," he said." But there's more to do. "Huge progress doesn't mean no problems yet to solve. We know more than we knew then. The only thing that most of the public sees is what's left to be done, and that's okay, because there's still lots to be done. I guess when you come to the end of your career, you want to look back and say, 'Was it worth it?' And my answer is yeah, it was definitely worth it."[4]

Otis Brown, who was at the center of the Virginia response to Kepone, reflected on the politics of managing a crisis. He felt that the Kepone response was much less politicized than a similar situation might be today. For Brown, the Godwin administration handled the situation, particularly the river closure, based on data, not partisanship. "But I think the key part of it... is [that] you make decisions based upon the data that you have. You take action based upon what appears to be the issue and... can be handled now. But you leave the political side totally off the table. And right now, that's hard to do because we have such an enormous lack of trust in the governmental process." At the time of his interview, Brown was thinking about the situation in Flint, Michigan, where lead entered the city's drinking water supply. "There are going to be some other Kepones. No question about it. The Flint, Michigan, situation. That's a little different situation. Apparently what's occurred is more political than anything else. The good part about this one, Kepone, [is that] it was never political. And I can't stress that enough."[5] By "political," Brown may have meant "partisan." Certainly the process of investigation and regulation of toxic substances is political.

Interviewed during the coronavirus pandemic, Bill Cummings drew on the legal significance of the Kepone cases.

> I was glad to see that people were paying attention to it from a safety standpoint. I tend to think that people forget bad things. I'm not too sure that people look back now and remember. That's forty-five years, for gosh sake. And so now we have this crazy virus, which has overshadowed everything we've ever had for the last fifty years. But cases like that need to be brought [up] and publicized when violations take place so that we can keep some control over industry. But . . . look at a situation . . . where the industrial giants are saying, "We don't care if we're sending out poison to this system. It's just an annoyance to us to have to worry about putting filters in and things like that." With that kind of attitude, which seems to be continuing to prevail among some industries, the public be damned. It's a very serious matter. It still is today.[6]

Dr. John Taylor, who treated some of the workers, reflected on the health aspects of Kepone contamination. He sees the lax safety concerns at Life Science as key. "In retrospect, this should never have happened. Money is the root of all evil. And I think they were so excited about making all this money. The workers were too, because they were making a lot. They just didn't bother to think clearly . . . "what the hell is this doing to us?" And they got the Kepone shakes. It would have been so easy to prevent this. All they had to do was get consultation with an industrial hygienist, and one visit would probably be enough to straighten them out with what they had to do. But that didn't happen." As to shutting down the river? "I think that was a little bit of an overkill. I don't think it was necessary. You could advise people not to eat a lot of fish, and that was probably all you needed to do. Fish is not a big part of most peoples' diet anyway."[7]

Retrospective newspaper stories included reflections from watermen and others in the seafood industry. As Taylor did, most believed it was overkill to close to river, reflecting the arguments raised at the time by watermen, some scientists, and political leaders like Herb Bateman. According to one news story from 1995, for watermen Kepone was "an old nightmare they would rather forget." One seafood broker claimed, "Kepone basically never hurt anything. The only lasting effect is when you guys bring it up."[8] In another story from 1995, an owner of a marina and motel on the James whose business suffered under the fishing ban was highly critical. "It was stupid, it really was. If you listen to the so-called experts, you realize how much of a thing you have to eat to really get contaminated with it." She didn't worry about the effects. "If I had been exposed to heavy doses of it, yeah, I would have been concerned." A waterman from James City County said that, on the good side, it made people aware: "People had to clean their act up. The James is cleaner than it's ever been in my lifetime."[9]

The reflections of Frank Arrigo and Tom Fitzgerald, both of whom suffered from Kepone poisoning, range in scope from the personal to the wider field of social and political relations. "I think about it," said Fitzgerald. "It probably crosses my mind, at some point, every day, but it's a long, long time ago, and it's just part of the background noise and has been for quite a while."

> I expect management people to have a management perspective. I expect working-class heroes to have a working-class perspective, and . . . it's just two different approaches. There's no way that everybody can agree. For me, maybe the most important thing that came out of it was that it was one of several lawsuits within a few years of each other that were responsible for people, at least briefly, paying closer attention. As a national disaster, I think it worked out as well as it could have. Something had to happen to make it very obvious that there was a problem, not only here, but perhaps in this industry, and enough good people paid attention to establish the kind of regulations that help interfere with that sort of blind greed. That's probably the best that you can hope for.[10]

Arrigo reflected on broken trust. "Well, you have to trust yourself to trust the people that you're working for. Don't take anything they say for granted. I mean, because they told us, like the main chemist, you can eat a cupful of this stuff, it won't hurt you. Which was an outright lie. That's all it was." He also reflected at length on the possible health effects from Kepone. Along with contracting scleroderma (and his siblings too), he recalled one of his coworkers, "thirty years older," who developed stomach cancer. "I'm not saying that [Kepone] absolutely caused the cancer, but you can't rule it out either way."[11]

Others are concerned about the long-term health risks. Jane Tarter recalled, "In 1975 I was pregnant with my second child, and my husband and my older son and I went camping with several other couples at a campground on the Chickahominy River." She and others ate fish from the river. Days later she learned that it had been closed because of Kepone. Sixteen years later, both she and her younger son were diagnosed with cancer. Was it Kepone? "I think it's very interesting, and is it a coincidence that I and my child-in-utero came down with cancer sixteen years after exposure to Kepone? There's no way we could prove it, we don't have any tissue samples surviving for anyone to do any diagnosis, but I think it's very suspicious."[12]

Linda Crowe is a writer and former official with the Nature Conservancy in Virginia. She grew up in Hopewell and sees Kepone as a cautionary tale. She remains suspicious about the fact that Kepone is still in the river, perhaps from a leak in one or more of the containment facilities on the Allied property. She's also concerned about the long-term health effects of Kepone on the workers, and cancer in Hopewell more generally. She reached out on Facebook to see if others in her group had anything to relate. The replies she received only added to her concerns. One person wrote: "I had a friend who lost her baby from being exposed to 'Second Hand Kepone.'" Another said: "I had a family member who worked at Life Science and was left sterile before he ever fathered a child. He has had cancer." Others replied that cancer was a key issue in general for Hopewell, as they reported various family members and friends who died of cancer. The data reveal the continued risks associated with living in an industrial city.

There is no definitive link to Kepone in these cases of illness or the health effects suffered by Jane Tarter, Frank Arrigo, and their families. But it can't be ruled out. What is certain is that these memories speak to how the residues of Kepone remain in the minds of many, whether as a point of reflection on policy as it was for Paylor and Maroon, personal memory, or a fear of Kepone's long-term health effects. For some, it's an example of government overreach, born out of fear and uncertainty. These memories and reflections reveal the dynamic and complex set of relationships embedded in Kepone, psychological residues of the crisis that will continue.

Of course, substances like Kepone can be intractable in the environment as well. The residents of the French Caribbean know this all too well. Those in regulatory organizations recognized the difficulty of being able to somehow manage them. The federal Council on Environmental Quality readily admitted in 1978, "Bringing toxic substances under control is more easily said than done. The number of chemical substances and the size of the chemical industry suggest the magnitude of the task."[13] There were some sixty-two thousand chemicals in commerce first listed through the TSCA; as of 2020 there were some eighty-six thousand. The TSCA proved inadequate to test or prevent toxic substances from entering the environment and people's bodies. These substances evade controls meant to contain them. When enacted, the TSCA allowed the sixty-two thousand chemicals to be grandfathered in without testing. As historian Brooks Flippen notes, the National Academy of Sciences acknowledged that the EPA "overly emphasized acute environmental effects at the expense of long-term consequences."[14] Through at least 2016, while more than twenty thousand new chemicals entered the market, only five were removed.[15] Congress updated the law with the Frank R. Lautenberg Chemical Safety for the 21st Century Act, signed by President Obama in June 2016. Unlike the TSCA, the new legislation does a number of things to improve safety, including mandating reviews of chemicals in commerce, a safety finding before they can enter the market, a health-based (as opposed to cost-benefit) safety standard, and explicit protections for vulnerable populations such as children and pregnant women. It also limited companies' abilities to claim information as private.[16]

In Virginia, under federal law the state is required to report on toxic substances released through the Toxic Release Inventory (TRI). Most releases are allowed under federal or state permits. While useful, this inventory does not account for all industrial pollution since it does not include all toxic chemicals or all sectors of the economy. As of 2020, amid ongoing political battles to either strengthen or weaken these regulations, approximately 28.3 million pounds of toxic chemicals were reported released on-site in Virginia, 14.6 million in the air, 12.2 in the water, and 1.4 million in the land. Most of the reported releases occurred under a federal or state permit program designed to protect human health and the environment. The total reported releases have dropped since 2003, when the total reached over 75 million pounds. In recent years, though, the total has remained relatively constant, around 28 million pounds between 2018 and 2020.[17] With all that might have come from Kepone, Virginia ranks seventh among fifty-six ranked states and territories on risk screening for toxic substances.[18]

Kepone was but one of the many pesticides created to assert control over insects. They are effective, yet, if that effectiveness wears off, the answer is not to abandon pesticides, but to find another, more effective formulation. Over time, insects develop resistance to chemicals. Out of a group of insects doused with poison, most die, but a few evolutionary outliers survive. When they breed, their offspring are resistant. This in turn generates the pressure and need for a new pesticide, and so it goes. As Rachel Carson wrote: "Darwin himself could scarcely have found a better example of the operation of natural selection than is provided by the way the mechanism of resistance operates."[19] Agrochemical corporations also have a financial interest in the need for newer insecticides; they operate under a form of planned obsolescence applied to agricultural products. Corporate leaders and scientists working with pesticides know that weeds, pests, and fungi harmful to crops, animals, and people will develop resistance to a particular chemical. They develop new chemicals and urge users to purchase those to defeat the newer "superpests." The more pesticides are used, the more they are needed.[20]

Within this global expansion of pesticides, Allied and Life Science were ready to supply the world with Kepone. While Virginia's Kepone crisis stemmed from the manufacture of the pesticide, tracing the chemical relations leads outside Hopewell to the public health crisis ongoing in Guadeloupe and Martinique. Hopewell and the French Caribbean were and remain landscapes of exposure, connected by Kepone. They are sites of waste and toxicity, residues that are "foundational categories of knowledge for the Anthropocene." They also remind us that "the Anthropocene may be planetary, but it is not uniform." Communities on the front lines of toxic exposure, whether in production or use, experience these large-scale transformations differently.[21] While there are important differences, Hopewell and the islands of Guadeloupe and Martinique both suffered from contamination and continue to deal with the aftereffects.

This is but one story of Kepone; there are many others ready to be told. Writing the stories of these places and events remains paramount, and within them giving voice to the multiplicity of memory and working to reveal the complexity of chemical relations. Even as some have wished to forget Kepone, that would be a mistake. In his work *Let Us Now Praise Famous Gullies*, Paul Sutter refers to "ecology of erasure," the "ways in which environmental forces have, with much human encouragement, obscured the South's history of human-induced erosion from view."[22] While his work focuses on soil erosion, the concept of erasure is a powerful idea, reminding us of how landscapes might obscure as much as they reveal, the result of the combined forces of cultural and natural processes. Because the rivers and waterways around Hopewell are healthier than they were in the 1970s, it becomes easy to erase or simply dismiss what happened. It's easy to obscure Kepone as it remains buried in landfills and river sediments, its records housed in boxes sitting in warehouses or archives, its experiences in the minds and bodies of those who lived it, memories only revealed (sometime reluctantly) when someone asks. Kepone is but one example of how chemical relations that shape our lives and transcend our local worlds are integral to our understanding of the humanity in nature and the nature in humanity.

NOTES

Introduction

1. Frank Arrigo, interview with the author, March 18, 2017.

2. Barbara J. Kuhn, "A Look at the Kepone Incident in Retrospect," PhD diss., University of Cincinnati, 1979, 2; and Environmental Protection Agency, *Reviews of the Environmental Effects of Pollutants: Mirex and Kepone* (Cincinnati: U.S. Environmental Protection Agency, 1978), 39–40.

3. Memo, Virginia Attorney General to State Water Control Board, January 20, 1976, box 9, "Kepone" folder, General Correspondence, 1976, Record Group 3, Executive Papers of Governor Mills E. Godwin Jr., 1974–1978, Library of Virginia, Richmond.

4. Jeanie LeNoir Langford, interview with the author, October 4, 2016.

5. C. Brian Kelly, "Kepone," in *Who's Poisoning America: Corporate Polluters and Their Victims in the Chemical Age*, ed. Ralph Nader, Ronald Brownstein, and John Richard (San Francisco: Sierra Club, 1981), 105–6; and Kuhn, "A Look at the Kepone Incident," 25–26.

6. "Memorandum in Support of Motion for Summary Judgment, Statement of Facts," September 17, 1979, box 1183, 77-3, William P. Moore v. Allied Chemical Corporation, CA 77-0379-R, file no. 6, Virginia-Eastern District, Records of District Courts of the United States, RG 21, NARA-Philadelphia.

7. Samuel S. Epstein, "Kepone-Hazard Evaluation," *Science of the Total Environment* 9, no. 1 (1978): 1–62.

8. Donald (Tom) Fitzgerald, interview with the author, December 5, 2016.

9. Y. M. Cabidoche et al., "Long-Term Pollution by Chlordecone of Tropical Volcanic Soils in the French West Indies: A Simple Leaching Model Accounts for Current Residue," *Environmental Pollution* 157, no. 5 (2009): 1697–1705.

10. Soraya Boudia et al., "Residues: Rethinking Chemical Environments," *Engaging Science, Technology, and Society* 4 (2018): 165–78, https://doi.org/10.17351/ests2018.245; Gabrielle Hecht, "Residue," *Somatosphere* (blog), January 8, 2018, http://somatosphere.net/2018/residue.html; and Gabrielle Hecht, "Interscalar Vehicles for an African Anthropocene: On Waste, Temporality, and Violence," *Cultural Anthropology* 33, no. 1 (2018): 109–41, https://doi.org/10.14506/ca33.1.05.

11. Elizabeth D. Blum, *Love Canal Revisited: Race, Class, and Gender in Environmental Activism* (Lawrence: University Press of Kansas, 2008); Robert D. Bullard, *Dumping in Dixie: Race, Class, and Environmental Quality*, 3rd ed. (Boulder, Colo.: Routledge, 2000); Steve Lerner and Phil Brown, *Sacrifice Zones: The Front Lines of Toxic Chemical Exposure in the United States* (Cambridge, Mass.: MIT Press, 2012); Steve Lerner and Robert D. Bullard, *Diamond: A Struggle for Environmental Justice in Louisiana's Chemical Corridor* (Cambridge, Mass.: MIT Press, 2006); Richard S. Newman, *Love Canal: A Toxic History from Colonial Times to the Present* (Oxford: Oxford University Press, 2019); and Ellen Griffith Spears, *Bap-*

tized in *PCBs: Race, Pollution, and Justice in an All-American Town*, New Directions in Southern Studies (Chapel Hill: University of North Carolina Press, 2014).

12. Brian Allen Drake, *Loving Nature, Fearing the State: Environmentalism and Antigovernment Politics before Reagan* (Seattle: University of Washington Press, 2013); A. E. Dick Howard, "Constitutional Revision: Virginia and the Nation," *University of Richmond Law Review* 9, no. 1 (1974): 1–49; Christine Keiner, *The Oyster Question: Scientists, Watermen, and the Maryland Chesapeake Bay since 1880* (Athens: University of Georgia Press, 2009); and Chris Wilhelm, "Conservatives in the Everglades: Sun Belt Environmentalism and the Creation of Everglades National Park," *Journal of Southern History* 82, no. 4 (2016): 823–54.

13. Paul S. Sutter and Christopher J. Manganiello, eds., *Environmental History and the American South: A Reader* (Athens: University of Georgia Press, 2009), 19.

14. Vanessa Agard-Jones, "Bodies in the System," *Small Axe* 17, no. 3 (2013): 182–92.

15. David Kinkela, *DDT and the American Century: Global Health, Environmental Politics, and the Pesticide That Changed the World* (Chapel Hill: University of North Carolina Press, 2011), 7; and Paul S. Sutter, "Tropical Conquest and the Rise of the Environmental Management State: The Case of U.S. Sanitary Efforts in Panama," in *Colonial Crucible: Empire in the Making of the Modern American State*, ed. Alfred W. McCoy and Francisco A. Scarano (Madison: University of Wisconsin Press, 2009), 317–26.

16. Mona Domosh, *American Commodities in an Age of Empire* (New York: Routledge, 2006), 3; and Nick Cullather, *The Hungry World: America's Cold War Battle against Poverty in Asia* (Cambridge, Mass.: Harvard University Press, 2013).

17. Joshua Blu Buhs, *The Fire Ant Wars: Nature, Science, and Public Policy in Twentieth-Century America* (Chicago: University of Chicago Press, 2004); and James C. Giesen, *Boll Weevil Blues: Cotton, Myth, and Power in the American South* (Chicago: University of Chicago Press, 2011).

18. Edmund Russell, *War and Nature: Fighting Humans and Insects with Chemicals from World War I to Silent Spring* (Cambridge: Cambridge University Press, 2001); and David Zierler, *The Invention of Ecocide: Agent Orange, Vietnam, and the Scientists Who Changed the Way We Think about the Environment* (Athens: University of Georgia Press, 2011).

19. Pete Daniel, *Toxic Drift: Pesticides and Health in the Post–World War II South* (Baton Rouge: Louisiana State University Press, 2007); and Michelle Mart, *Pesticides, A Love Story: America's Enduring Embrace of Dangerous Chemicals* (Lawrence: University Press of Kansas, 2015).

20. Mark Hersey and James C. Giesen, "The New South and the Natural World," in *Interpreting American History: The New South*, ed. James Humphreys (Kent: Kent State University Press, 2017), 241–66; and Adam Rome, "What Really Matters in History: Environmental Perspectives on Modern America," *Environmental History* 7, no. 2 (2002): 303–18.

21. Andrew C. Isenberg, ed., *The Oxford Handbook of Environmental History* (Oxford: Oxford University Press, 2014), 716.

22. Lerner and Brown, *Sacrifice Zones*.

23. See Matthew D. Lassiter and Joseph Crespino, eds., *The Myth of Southern Exceptionalism* (New York: Oxford University Press, 2010).

24. Kelly E. Happe, Jenell Johnson, and Marina Levina, eds., *Biocitizenship: The Politics of Bodies, Governance, and Power* (New York: New York University Press, 2018); Adriana Petryna, *Life Exposed: Biological Citizens after Chernobyl* (Princeton: Princeton University Press, 2002);

and Nikolas Rose, *The Politics of Life Itself: Biomedicine, Power, and Subjectivity in the Twenty-First Century* (Princeton: Princeton University Press, 2007).

25. Michelle Brattain, *The Politics of Whiteness: Race, Workers, and Culture in the Modern South* (Athens: University of Georgia Press, 2004); and Jonathan Metzl, *Dying of Whiteness: How the Politics of Racial Resentment Is Killing America's Heartland* (New York: Basic Books, 2019).

26. Christof Mauch and Thomas Zeller, eds., *Rivers in History: Perspectives on Waterways in Europe and North America* (Pittsburgh: University of Pittsburgh Press, 2008); and Richard White, *The Organic Machine: The Remaking of the Columbia River* (New York: Hill & Wang, 1996).

27. Keiner, *The Oyster Question*; and Michael Paolisso, "Blue Crabs and Controversy on the Chesapeake Bay: A Cultural Model for Understanding Watermen's Reasoning about Blue Crab Management," *Human Organization* 61, no. 3 (2002): 226–39.

28. Brinda Sarathy, Vivien Hamilton, and Janet Farrell Brodie, eds., *Inevitably Toxic: Historical Perspectives on Contamination, Exposure, and Expertise* (Pittsburgh: University of Pittsburgh Press, 2018), 8.

29. Michelle Murphy, *Sick Building Syndrome and the Problem of Uncertainty: Environmental Politics, Technoscience, and Women Workers* (Durham : Duke University Press, 2006).

30. Gregg Mitman, Michelle Murphy, and Christopher Sellers, "Introduction: A Cloud over History," *Osiris* 19 (2004): 1–17, https://doi.org/10.1086/649391; and Robert N. Proctor and Londa Schiebinger, eds., *Agnotology: The Making and Unmaking of Ignorance* (Stanford, Calif.: Stanford University Press, 2008).

Chapter 1. The James River before Kepone

1. Christof Mauch and Thomas Zeller, eds., *Rivers in History: Perspectives on Waterways in Europe and North America* (Pittsburgh: University of Pittsburgh Press, 2008), 7; and Richard White, *The Organic Machine: The Remaking of the Columbia River* (New York: Hill & Wang, 1996).

2. Larry S. Chowning, *Harvesting the Chesapeake: Tools and Traditions* (Centreville, Md.: Tidewater, 1990); and *Kepone Contamination: Hearings before the Subcommittee on Agricultural Research and General Legislation of the Committee on Agriculture and Forestry, U.S. Senate*, 94th Cong., 2nd sess. (1976), http://hdl.handle.net/2027/mdp.39015071090206.

3. *Report on the Economic Potential and Management of Virginia's Seafood Industry to the Governor and General Assembly of Virginia* (Richmond: Commonwealth of Virginia, 1984), 2.

4. Blair Niles, *The James: From Iron Gate to Sea*, 2nd ed. (New York: Farrar & Rinehart, 1945).

5. Committee on Contaminated Marine Sediments et al., *Contaminated Marine Sediments: Assessment and Remediation* (Washington, D.C.: National Academies Press, 1989), 412; and Federal Highway Administration, "Route 31, James River Crossing from Route 10 to Route 5, Charles City/James City/Surry Counties: Environmental Impact Statement," 1989, 31.

6. Not part of the Kepone story directly was another anadromous fish, the Atlantic sturgeon (*Acipenser oxyrinchus*). It was once a critical species in the river. It served as a main source of food for Indians and English settlers, and the eggs were a valuable source of caviar. It supported a valuable fishing industry, but that declined rapidly by the early twentieth century.

Already threatened by 1975, it was a rare sight in the James then. However, recent developments in habitat restoration have seen the fish return to the river. See "Atlantic Sturgeon," U.S. Fish and Wildlife Service," accessed January 30, 2022, https://www.fws.gov/fisheries/freshwater-fish-of-america/atlantic_sturgeon.html.

7. "American Eel," fact sheet, U.S. Fish and Wildlife Service, October 2015, accessed May 20, 2022, https://www.fws.gov/northeast/americaneel/pdf/American_Eel_factsheet_2015.pdf.

8. Patrick S. Lookabaugh and Paul L. Angermeier, "Diet Patterns of American Eel, Anguilla Rostrata, in the James River Drainage, Virginia," *Journal of Freshwater Ecology* 7, no. 4 (1992): 425–31.

9. Atlantic States Marine Fisheries Commission, *Interstate Fisheries Management Plan for the Striped Bass* (Atlantic States Marine Fisheries Commission, 1981), http://www.asmfc.org/uploads/file/1981FMP.pdf.

10. John McPhee, *The Founding Fish* (New York: Farrar, Straus & Giroux, 2002), 14.

11. Ryan W. Schlosser et al., "Ecological Role of Blue Catfish in Chesapeake Bay Communities and Implications for Management," *VIMS Books and Book Chapters* 10 (2011): 369–82.

12. "Crassostrea Virginica," Indian River Lagoon Species Inventory, accessed January 30, 2022, https://irlspecies.org/taxa/index.php?taxon=4347&taxauthid=1.

13. Eastern Oyster Biological Review Team, "Status Review of the Eastern Oyster (Crassostrea Virginica): Report to the National Marine Fisheries Service, Northeast Regional Office," February 16, 2007.

14. "Blue Crabs," Chesapeake Bay Foundation, accessed January 30, 2022, https://www.cbf.org/about-the-bay/more-than-just-the-bay/chesapeake-wildlife/blue-crabs/index.html.

15. John Olney and John Hoenig, *Monitoring Relative Abundance of American Shad in Virginia's Rivers* (Gloucester Point: Virginia Institute of Marine Science, College of William & Mary, 2001); and Virginia Commission on Fisheries, "Report of the Fish Commissioners to the Governor of Virginia," 1872, https://babel.hathitrust.org/cgi/pt?id=hvd.hwhphx;view=1up;seq=9.

16. Kate Livie, *Chesapeake Oysters: The Bay's Foundation and Future* (Charleston, S. C.: History Press, 2015), 12.

17. David M. Schulte, "History of the Virginia Oyster Fishery, Chesapeake Bay, USA," *Frontiers in Marine Science* 4 (2017), https://doi.org/10.3389/fmars.2017.00127.

18. James Tice Moore, "Gunfire on the Chesapeake: Governor Cameron and the Oyster Pirates, 1882–1885," *Virginia Magazine of History and Biography* 90, no. 3 (1982): 367–77; and John R. Wennersten, *The Oyster Wars of Chesapeake Bay* (Centreville, Md.: Tidewater, 1981).

19. L. Eugene Cronin, "Early Days of Crabbing and a Brief History of the Chesapeake Bay," *Journal of Shellfish Research* 17, no. 2 (1998): 379–81.

20. Ann Woodlief, "From Energy to Entropy," chap. 15 in *River Time: The Way of the James* (Chapel Hill, N.C.: Algonquin Books of Chapel Hill, 1985), accessed May 23, 2018, https://archive.vcu.edu/english/engweb/Rivertime/chp15.htm.

21. Claudia Copeland, *Clean Water Act: A Summary of the Law* (Washington, D.C.: Congressional Research Service, 2016).

22. "Shenandoah National Park," *Encyclopedia Virginia*, accessed May 22, 2018, https://www.encyclopediavirginia.org/entries/shenandoah-national-park; and Sara M. Gregg, *Managing the Mountains: Land Use Planning, the New Deal, and the Creation of a Federal Landscape in Appalachia* (New Haven, Conn.: Yale University Press, 2010), 115.

23. Joint Legislative Audit and Review Commission, Virginia General Assembly, *Program

Evaluation: Water Resource Management in Virginia (Richmond: The Commission, 1976), http://jlarc.virginia.gov/pdfs/reports/Rpt05.pdf.

24. Ronald L. Heinemann et al., *Old Dominion, New Commonwealth: A History of Virginia, 1607–2007* (Charlottesville: University of Virginia Press, 2008), 299.

25. Heinemann et al., *Old Dominion, New Commonwealth*, 306–9.

26. Virginia Academy of Science, *The James River Basin: Past, Present, and Future* (Richmond: Virginia Academy of Science, 1950), xxvii.

27. William J. Hargis, "The Virginia Institute of Marine Science: Virginia's Official Marine Science, Engineering and Advisory Program," W&M ScholarWorks, 1974, http://dx.doi.org/doi:10.21220/m2-yqn8-cg54.

28. Fred G. Kern, "Minchinia Nelsoni (MSX) Disease of the American Oyster," *Marine Fisheries Review* 38, no. 10 (1976).

29. Morris L. Brehmer and Samuel O. Haltiwanger, "A Biological and Chemical Study of the Tidal James River," W&M ScholarWorks, May 19 1966), accessed May 20, 2022, https://scholarworks.wm.edu/cgi/viewcontent.cgi?article=1927&context=reports.

30. League of Women Voters of Virginia, James River Basin: A Progress Report (Williamsburg, Va.: n.p., 1966).

31. See "Guide to the Fitzgerald Bemiss Papers," Virginia Historical Society, http://ead.lib.virginia.edu/vivaxtf/view?docId=vhs/vi00002.xml;query=bemiss;brand=default.

32. Gregory L. Poe, *The Evolution of Federal Water Pollution Control Policies* (Ithaca, N.Y.: Dept. of Agricultural, Resource, and Managerial Economics, Cornell University, 1995).

33. Edmund S. Muskie, "NEPA to CERCLA: The Clean Air Act: A Commitment to Public Health," CleanAir Trust, accessed October 16, 2018, http://www.cleanairtrust.org/nepa2cercla.html (originally published in the January/February 1990 issue of *Environmental Forum*). See also William B. Spong, *A Man for Today: A Collection of Speeches* (N.p., 1971), and "William Belser Jr. Spong (1920–1997)," *Encyclopedia Virginia*, accessed October 16, 2018, https://www.encyclopediavirginia.org/Spong_William_Belser_Jr_1920-1997#start_entry.

34. Brian Allen Drake, *Loving Nature, Fearing the State: Environmentalism and Antigovernment Politics Before Reagan* (Seattle: University of Washington Press, 2013), 183.

35. A. E. Dick Howard, "Constitutional Revision: Virginia and the Nation," *University of Richmond Law Review* 9, no. 1 (1974): 1–49; Christine Keiner, *The Oyster Question: Scientists, Watermen, and the Maryland Chesapeake Bay since 1880* (Athens: University of Georgia Press, 2009); Geoffrey Skelley, "Virginia's Redistricting History: What's Past Is Prologue," University of Virginia Center for Politics, June 18, 2015, https://centerforpolitics.org/crystalball/articles/virginias-redistricting-history-whats-past-is-prologue; and "Constitution of Virginia, Article XI. Conservation," Virginia State Law Portal, accessed May 18, 2021, https://law.lis.virginia.gov/constitutionexpand/article11/.

36. Chris Wilhelm, "Conservatives in the Everglades: Sun Belt Environmentalism and the Creation of Everglades National Park," *Journal of Southern History* 82, no. 4 (2016): 829.

37. "Higher Standards for Cleaner Water," *Virginian Pilot*, December 9, 1969, A14, box 5, "Environment: Water Pollution, 1969–1970" folder, Fitzgerald Bemiss Papers, Virginia Museum of History & Culture, Richmond.

38. James L. Bugg Jr., "Mills Edwin Godwin, Jr.: A Man for All Seasons," in James Tice Moore and Edward Younger, *The Governors of Virginia, 1860–1978* (Charlottesville: University of Virginia Press, 1982). See also "Mills E. Godwin," Encyclopedia Virginia, https://www.encyclopediavirginia.org/Godwin_Mills_E_1914-1999#start_entry.

39. Linwood Holton, *Opportunity Time* (Charlottesville: University of Virginia Press, 2008), 99.

40. Council on the Environment, *Our Common Wealth: Virginians View Their Environment* (Richmond, Va.: The Council, 1971).

41. Gerald P. McCarthy, interview with the author, October 25, 2016.

42. "A Conversation with the Times-Dispatch: Linwood Holton," *Richmond Times-Dispatch*, March 30, 2008.

43. Governor's Council on the Environment, *Annual Report of the Governor's Council on the Environment* (Richmond, Va.: The Council, 1971); and Virginia, Council on the Environment, *The State of Virginia's Environment: Annual Report of the Council on the Environment* (Richmond, Va.: The Council, 1973).

44. James S. Burton, *History of the State Water Resources Research Institute Program* (Washington, D.C.: U.S. Geological Survey, 1984), https://pubs.usgs.gov/of/1984/0736/report.pdf.

Chapter 2. Hopewell, Allied, and the Beetle Battles

1. Clint Strong, interview with the author, November 11, 2016.

2. Linda Crowe, email to author, 2017.

3. Jeanie LeNoir Langford, interview with the author, October 4, 2016.

4. Wayne Walton, interview with the author, November 4, 2016.

5. Mark Haley, interview with the author, November 4, 2016.

6. Gregg Mitman, Michelle Murphy, and Christopher Sellers, "Introduction: A Cloud over History," *Osiris* 19 (2004): 12, https://doi.org/10.1086/649391.

7. Michelle Mart, *Pesticides, a Love Story: America's Enduring Embrace of Dangerous Chemicals* (Lawrence: University Press of Kansas, 2015), 6.

8. Virginia Writers' Project, *A Guide to Prince George and Hopewell* (n.p.: Hopewell News, 1939).

9. "Capt. Francis Eppes Making Friends with the Appomattox Indians," accessed July 1, 2018, https://postalmuseum.si.edu/exhibition/indians-at-the-post-office-murals-encounter/capt-francis-eppes-making-friends-with-the; Virginia Writers' Project, *A Guide to Prince George and Hopewell*; and Marie Tyler-McGraw, *Slavery and the Underground Railroad at the Eppes Plantations, Petersburg National Battlefield* (Petersburg, Va.: National Park Service, 2005), https://www.nps.gov/parkhistory/online_books/pete/ugrr.pdf.

10. Tyler-McGraw, "Slavery and the Underground Railroad"; and "Petersburg National Battlefield," National Park Service, accessed July 1, 2018, https://www.nps.gov/pete/learn/historyculture/richard-eppes.htm.

11. Benjamin R. Cohen, *Notes from the Ground: Science, Soil, and Society in the American Countryside* (New Haven, Conn.: Yale University Press, 2014), 6; and Shearer Davis Bowman, "Conditional Unionism and Slavery in Virginia, 1860–1861: The Case of Dr. Richard Eppes," *Virginia Magazine of History and Biography* 96, no. 1 (1988): 31–54.

12. Laurie Rizzo, "DuPont Company Museum Collection," finding aid, Hagley Museum & Library, https://findingaids.hagley.org/repositories/2/resources/16.

13. Gilbert C. Fite, "Southern Agriculture since the Civil War: An Overview," *Agricultural History* 53, no. 1 (1979): 3–21; and John Beck, Wendy Jean Frandsen, and Aaron Randall, *Southern Culture: An Introduction*, 3rd ed. (Durham: Carolina Academic, 2012).

14. E. I. Du Pont de Nemours, *Hand Book of Explosives for Farmers, Planters, Ranchers*

(Wilmington, Del.: E. J. Du Pont de Nemours Powder, 1910), http://archive.org/details/handbookofexplosoodupo.

15. *Hardware Dealers' Magazine*, June 1919.

16. Alfred Dupont Chandler and Stephen Salsbury, *Pierre S. Du Pont and the Making of the Modern Corporation* (Washington, D.C.: Beard, 2000), 423.

17. Roy MacLeod and Jeffrey A. Johnson, *Frontline and Factory: Comparative Perspectives on the Chemical Industry at War, 1914–1924* (Dordrecht, Netherlands: Springer, 2007), 173.

18. Raymond Perkinson, "Hopewell: Boom Town to Solid City," *Commonwealth*, September 1952, 23–26.

19. According to Hopewell local historian Jeanie Langford, Archie Vincent Carey worked for Tubize until the plant closed in 1934. He then opened his own business as a refrigerator technician and appliance repairman; he passed away in 1965. *Pictorial History of Hopewell, Virginia* is a reprint of newspaper articles about the city, sponsored by local businesses to put the city in the best light.

20. A. V. Carey, *Pictorial History of Hopewell, Virginia: Illustrating the Development of the Eighth Wonder of the World* (Hopewell, Va.: A. V. Carey, 1916).

21. House Select Committee on Expenditures in the War Department, *War Expenditures: Hearings before Subcommittee No. 5 (Ordnance) of the Select Committee on Expenditures in the War Department, House of Representatives, Sixty-Sixth Congress, First–[Third] Session, on War Expenditures*, 66th Cong., 2nd and 3rd sess. (1919–1921), http://hdl.handle.net/2027/uc1.c061046012.

22. Nancy Carter Crump, "Hopewell during World War I: 'The Toughest Town North of Hell,'" *Virginia Cavalcade*, Summer 1981, 38–47.

23. Crump, "Hopewell during World War I," 38.

24. Carey, *Pictorial History of Hopewell*, 4.

25. R. W. Fitzgerald, The City That DuPont Built (Richmond, Va.: R. W. Fitzgerald, 1991).

26. Virginia Writers' Project, *Guide to Prince George and Hopewell*, 21.

27. E. I. du Pont de Nemours & Company, "An Answer to the Question of—Labor Troubles and Production," 1919, Hagley Digital Archives, Hagley Museum, http://digital.hagley.org/dpads_1803_00054.

28. Langford interview.

29. Gregg D. Kimball and Ron T. Curry, "On the Beach of Waikiki: Hopewell's Tubize Royal Hawaiian Orchestra," *Virginia Cavalcade* 51, no. 3 (2002): 113.

30. Janice Denton, interview with the author, December 2, 2016.

31. "Allied History," box 2, "Kepone–Allied Chemical" folder, C. Brian Kelly, C. Brian Kelly Papers, Albert and Shirley Small Special Collections Library, University of Virginia, Charlottesville.

32. Fred J. Emmerich, *Chemicals for American Progress* (New York: Newcomen Society, American Branch, 1956), 22–23.

33. Virginia Writers' Project, *Guide to Prince George and Hopewell*.

34. Kit Weigel, "Since Tubize, Hopewell Has Never Been the Same," *Hopewell News*, May 14, 1976.

35. "Allied Chemical," news clipping, Container 5, RG 131, Office of Alien Property, National Archives and Record Administration, Maryland (NARA-MD), College Park.

36. "History of the Hopewell, Virginia Chemical Plant," Honeywell/AlliedSignal Retirees' Home Page, accessed August 9, 2018, http://www.hon-area.org/hopewell.html.

37. Frank J. Soday, "The Chemical Industry Looks South," *Analysts Journal* 8, no. 5 (1952): 51–58; and Bruce J. Schulman, *From Cotton Belt to Sunbelt: Federal Policy, Economic Development, and the Transformation of the South 1938–1980*, rev. ed. (Durham: Duke University Press, 1994), 129.

38. *Prime Virginia Sites for Chemical and Water-Using Industries* (Richmond: Commonwealth of Virginia, Division of Industrial Development, 1969). See also Schulman, *Cotton Belt to Sunbelt*, especially chapters 6–7.

39. "Chlordecone," International Programme on Chemical Safety, Environmental Health Criteria 43, 1984, http://www.inchem.org/documents/ehc/ehc/ehc43.htm; and C. Brian Kelly, "Kepone," in *Who's Poisoning America: Corporate Polluters and Their Victims in the Chemical Age*, ed. Ralph Nader, Ronald Brownstein, and John Richard (San Francisco: Sierra Club, 1981), 89.

40. "Life Science Products Company," memorandum, State Attorney General to State Water Control Board, January 20, 1976, box 9, "Kepone" folder, Executive Papers of Governor Mills E. Godwin Jr., 1976, Library of Virginia, Richmond. See also *Toxicological Profile for Mirex and Chlordecone*, (U.S. Department of Health and Human Services, Agency for Toxic Substances and Disease Registry 1995), https://www.atsdr.cdc.gov/toxprofiles/tp66.pdf.

41. Mart, *Pesticides, a Love Story*, 6.

42. Russell, *War and Nature*, 189. See also David Kinkela, *DDT and the American Century: Global Health, Environmental Politics, and the Pesticide That Changed the World* (Chapel Hill: University of North Carolina Press, 2011).

43. Thomas L. Ilgen, "'Better Living through Chemistry': The Chemical Industry in the World Economy," *International Organization* 37, no. 4 (1983): 650.

44. Christopher J. Bosso, *Pesticides and Politics: The Life Cycle of a Public Issue*, Pitt Series in Policy and Institutional Studies (Pittsburgh: University of Pittsburgh Press, 1987), 70.

45. Charles Penrose, "The Newcomen Society in North America," *Public Historian* 3, no. 3 (1981): 129–31.

46. Emmerich, *Chemicals for American Progress*, 25.

47. *Pesticides and Public Policy* (Washington, D.C.: National Agricultural Chemicals Association, 1960), 5, http://hdl.handle.net/2027/wu.89064736069.

48. Pete Daniel, *Toxic Drift: Pesticides and Health in the Post-World War II South* (Baton Rouge: Louisiana State University Press, 2007), 2–4.

49. Jill Lindsey Harrison, *Pesticide Drift and the Pursuit of Environmental Justice* (Cambridge, Mass.: MIT Press, 2011), 86.

50. David Pimentel, "Green Revolution Agriculture and Chemical Hazards," *Science of the Total Environment* 188 (1996): 13.

51. Thomas Robertson, *The Malthusian Moment: Global Population Growth and the Birth of American Environmentalism* (New Brunswick, N.J.: Rutgers University Press, 2012), 1–2.

52. Robertson, *Malthusian Moment*, 142.

53. Carlos Kampmeier, "Food for All through World-wide Effort," *NAC News and Pesticide Review* 26, no. 2 (1967): 3–5.

54. See "Firms Exporting Products Banned as Risks in U.S.," *Washington Post*, February 25, 1980; and Ilgen, "Better Living through Chemistry," 659, 677.

55. James H. Lashomb and Richard Casagrande, eds., *Advances in Potato Pest Management* (Stroudsburg, Penn.: Hutchinson Ross, 1981), ix–x.

56. Carol Sheppard, "Benjamin Dann Walsh: Pioneer Entomologist and Proponent of Darwinian Theory," *Annual Review of Entomology* 49 (2004): 1–25.

57. Andrei Alyokhin et al., "Colorado Potato Beetle Resistance to Insecticides," *American Journal of Potato Research* 85, no. 6 (2008): 395–413.

58. Charles C. Mann, *1493: Uncovering the New World Columbus Created* (New York: Vintage, 2012), 288–89.

59. Benjamin D. Walsh, "The New Potato-bug, and Its Natural History," *Practical Entomologist* 1, no. 1 (1865).

60. Andrei Alyokhin, Silvia I. Rondon, and Yulin Gao, *Insect Pests of Potato: Global Perspectives on Biology and Management London: Academic Press, 2012), 13;* and Lucy Burns, "The Great Cold War Potato Beetle Battle," BBC, September 3, 2013, http://www.bbc.com/news/magazine-23929124.

61. Charles V. Riley, *Colorado Potato-beetle and the Other Insect Foes of the Potato in North America* (New York: Orange Judd, 1876), 50–59, 60.

62. Mann, *1493*, 235–36; W. Conner Sorensen, Edward H. Smith, and Janet R. Smith, *Charles Valentine Riley: Founder of Modern Entomology* (Tuscaloosa: University of Alabama Press, 2019), 52; and Robert L. Zimdahl, *Six Chemicals That Changed Agriculture* (Boston: Academic Press, 2015), 30.

63. John F. Clark, *Bugs and the Victorians* (New Haven, Conn.: Yale University Press, 2009), 151–53.

64. See Jorge E. Peña, Jennifer L. Sharp, and M. Wyoski, *Tropical Fruit Pests and Pollinators: Biology, Economic Importance, Natural Enemies and Control* (New York: CABI, 2002), 14–16; and Robert E. Woodruff and Thomas R. Fasulo, "Banana Root Borer, *Cosmopolites sordidus* (Germar) (Insecta: Coleoptera: Curculionidae)," Institute of Food and Agricultural Sciences Extension, University of Florida, 2015, https://edis.ifas.ufl.edu/pdffiles/IN/IN70600.pdf.

65. John Soluri, *Banana Cultures: Agriculture, Consumption, and Environmental Change in Honduras and the United States* (Austin: University of Texas Press, 2006), 6.

66. Soluri, *Banana Cultures*, 6.

67. Soluri, *Banana Cultures*, 10.

68. Soluri, *Banana Cultures*, 16.

69. Barbara M. Welch, *Survival by Association: Supply Management Landscapes of the Eastern Caribbean* (Montreal: McGill-Queen's University Press, 1998), 12.

70. The letter was included in the Senate hearings on Kepone, part of a subpoena of documents for that purpose.

71. *Report on Carcinogenesis Bioassay of Technical Grade Chlordecone (Kepone)* (Washington, D.C.: National Cancer Institute, 1976), https://web.archive.org/web/20111023024457/http://ntp.niehs.nih.gov/ntp/htdocs/LT_rpts/trChlordecone(kepone).pdf.

72. Kepone's fellow Allied Chemical cousin, Mirex, however, did find a strong market in the South against the fire ant. See Joshua Blu Buhs, *The Fire Ant Wars: Nature, Science, and Public Policy in Twentieth-Century America* (Chicago: University of Chicago Press, 2004), 142–43.

73. "Answers to Interrogatories," Allied Chemical to Hundley, Taylor, and Glass, November 10, 1975, box 1057, 75-3, "Dale F. Gilbert v. Allied Chemical Corporation," CA 75-0469-R, file 1, Virginia–Eastern District, Records of District Courts of the United States, RG 21, NARA-Philadelphia.

74. Memorandum, Allied Chemical, "Visit of Dr. Spiess," October 20, 1974, box 5, "Kepone-Rachel Carson Trust" folder, papers of C. Brian Kelly, Albert and Shirley Small Special Collections Library, University of Virginia, Charlottesville. As of 2021, the company is GECHEM.

75. Matthieu Fintz, *Éléments historiques sur l'arrivée du chlordécone en France entre 1968 et 1981* (Maisons-Alfort, France: Agence française de sécurité sanitaire de l'environnement et du travail, 2009), https://www.anses.fr/fr/system/files/SHS2009etPlanChlor01Ra.pdf?_x_tr_sl=fr&_x_tr_tl=en&_x_tr_hl=en&_x_tr_pto=sc.

76. Soluri, *Banana Cultures*, 210.

77. Pierre-Benoit Joly, *La saga du chlordécone aux Antilles français: Reconstruction chronologique 1968–2008* (Paris: Institut de Recherche sur les Fruits et Agrumes, 2010), 17–23, https://www.anses.fr/fr/system/files/SHS2010etInracolo1Ra.pdf.

78. Assemblée Nationale, "Impacts de l'utilisation de la chlordécone er des pesticides aux Antilles: Bilan et perspectives d'évolution," Sénat, https://www.senat.fr/rap/r08-487/r08-4873.html.

79. Buhs, *Fire Ant Wars*, 38.

80. James C. Giesen, *Boll Weevil Blues: Cotton, Myth, and Power in the American South* (Chicago: University of Chicago Press, 2011), xi, xii.

Chapter 3. The Kepone Shakes and a Poisoned River

1. "Bugless Sludge Put in Special Pit," *Hopewell News*, May 12, 1975.

2. Christopher Sellers, *Hazards of the Job* (Chapel Hill: University of North Carolina Press, 1997), 231.

3. Soraya Boudia et al., "Residues: Rethinking Chemical Environments," *Engaging Science, Technology, and Society* 4 (2018): 167, https://doi.org/10.17351/ests2018.245.

4. "William Moore," obituary, J. T. Morriss & Son Funeral Home & Cremation Service, July 1, 2004, https://www.jtmorriss.com/obituaries/William-Moore-000132/#!/Obituary.

5. See "Virgil Hundtofte, Director of Til Kuwait," Til Kuwait, http://tilkuwait.com/virgil-hundtofte.html. Mr. Hundtofte denied a request for an interview.

6. *Kepone Contamination: Hearings before the Subcommittee on Agricultural Research and General Legislation of the Committee on Agriculture and Forestry, U.S. Senate*, 94th Cong., 2nd sess. (1976), http://hdl.handle.net/2027/mdp.39015071090206, 47.

7. *Kepone Contamination*, 205.

8. Robert R. Merhige Jr. et al., "Panel Discussions: Allied Chemical, the Kepone Incident, and the Settlements: Twenty Years Later," *University of Richmond Law Review* 29, no. 493 (1995): 505; and P. Aarne Vesilind and Alastair S. Gunn, *Engineering, Ethics, and the Environment* (Cambridge, UK: Cambridge University Press, 1998), 144.

9. *Kepone Contamination*, 208; and Brian Kelly, "Kepone," in *Who's Poisoning America: Corporate Polluters and Their Victims in the Chemical Age*, ed. Ralph Nader, Ronald Brownstein, and John Richard (San Francisco: Sierra Club, 1981), 85.

10. *Kepone Contamination*, 206.

11. William Goldfarb, "Kepone: A Case Study," *Environmental Law* 8, no. 3 (1978): 649.

12. Goldfarb, "Kepone," 650; quotation: *Kepone Contamination*, 240. A copy of the contract appears in *Kepone Contamination*, 237–44.

13. *Kepone Contamination*, 166.

14. Samuel S. Epstein, "Kepone-Hazard Evaluation," *Science of the Total Environment*, no. 9 (1978): 17.

15. *Kepone Contamination*, 43, 142, 164.

16. *Kepone Contamination*, 257–60.

17. Gilbert's pay would be about $1,200 per week in 2020 dollars. "Employment History, Dale F. Gilbert," box 75-3, *Dale F. Gilbert, et al. vs. Allied Chemical Corporation*, CA 75-0469-R, file 3, NARA–Philadelphia.

18. *Oversight Hearings on the Occupational Safety and Health Act: Hearings before the Subcommittee on Manpower, Compensation, and Health and Safety of the Committee on Education and Labor, House of Representatives*, 94th Cong., 2nd sess. (1976), 47.

19. *Oversight Hearings on the Occupational Safety and Health Act*, 47.

20. Donald (Tom) Fitzgerald, interview with the author, December 5, 2016.

21. *Kepone Contamination*, 155.

22. Kit Weigel, interview with the author, October 4, 2016.

23. Harris quoted in "Kepone: Chemical Remains Fact of Life for Its Makers and Watermen," *Richmond Times-Dispatch*, June 9, 1985.

24. *Kepone Contamination*, 214.

25. Yinan Chou, email communication to the author, May 22, 2019.

26. Yinan Chou, interview with the author, May 14, 2019.

27. "The 2 Who Shut Down Poison Mill," *Newsday*, January 5, 1976.

28. Chou, email.

29. John Taylor, interview with the author, October 31, 2017.

30. Harold Waters, "Toxin Level Closes Firm in Hopewell," *Richmond Times-Dispatch*, August 12, 1975; and *Kepone Contamination*, 45. See also *Oversight Hearings on the Occupational Safety and Health Act*, 4.

31. James Kenley, interview with the author, October 11, 2016.

32. *Oversight Hearings on the Occupational Safety and Health Act*, 6; and *Kepone Contamination*, 71.

33. "Supporting Data for Extension of the Emergency Rule dated June 11, 1976," box 8, "State Board of Health Emergency Rules—Closing of James River 1975" folder, Virginia Department of Health, Office of the State Health Commissioner, Regulatory Case Files, 1975–1997, Library of Virginia, Richmond.

34. *Kepone Contamination*, 360–61.

35. *Kepone Contamination*, 35.

36. Yinan Chou, interview with the author, May 7, 2019.

37. *Oversight Hearings on the Occupational Safety and Health Act*, 28.

38. See *Kepone Contamination*, 146.

39. Fitzgerald, interview.

40. Kit Weigel, "EPA in Hopewell for Kepone Queries," *Hopewell News*, August 15, 1975, 1; and "Kepone Sends Officials on Door-to-Door Survey," *Hopewell News*, August 18, 1975, 1.

41. Robert H. Dreisbach, *Handbook of Poisoning: Diagnosis and Treatment*, 8th ed. (Los Altos, Calif.: Lange Medical, 1974).

42. *Oversight Hearings on the Occupational Safety and Health Act*, 22.

43. Treacy also served as assistant attorney general in Virginia; in that capacity he led the 1997 suit that fined Smithfield Foods $12.7 million for illegal discharges into the Pagan River. He then went on to change Smithfield's environmental program. See "Consensus Builder: Treacy Seeks Common Ground among Many Stakeholders," *Virginia Business*, September 29, 2017, https://www.virginiabusiness.com/article/consensus-builder1.

44. Dennis Treacy, interview with the author, March 8, 2017.

45. David Paylor, interview with the author, March 8, 2017.

46. "Fact Sheet on Kepone Levels Found in Environmental Samples from the Hopewell, VA Area," December 16, 1975, box 8, "State Board of Health Emergency Rules—Closing of James River 1975" folder, Virginia Department of Health, Office of the State Health Commissioner, Regulatory Case Files, 1975–1997, Library of Virginia, Richmond.

47. "Good News: Low Kepone in Blood," *Hopewell News*, March 8, 1976, 1.

48. "Virginia Seafood Council-Kepone Limits," legal memorandum, Thomas H. Truitt to Herbert Bateman, September 15, 1976, box 8, folder 21, series 93: Kepone, Herbert H. Bateman Papers, Swem Library, College of William & Mary, Williamsburg, Virginia.

49. Treacy, interview.

50. Paylor, interview.

51. Treacy, interview.

52. *Kepone Contamination*, 56.

53. *Oversight Hearings on the Occupational Safety and Health Act*, 257.

54. *Kepone Contamination*, 132.

55. International Agency for Research on Cancer, *Some Halogenated Hydrocarbons*, vol. 20, IARC Monographs on the Evaluation of the Carcinogenic Risk to Humans (Lyon, France: International Agency for Research on Cancer, 1979), https://monographs.iarc.fr/wp-content/uploads/2018/06/mono20.pdf.

56. *Oversight Hearings on the Occupational Safety and Health Act*, 278.

57. Otis Brown, interview with the author, June 24, 2018.

58. Paul Langan, *60 Minutes* segment "Kepone—1975 Pesticide Poisonings Allied Chemical," YouTube, https://www.youtube.com/watch?v=h2FAGvaOZHI.

59. Transcript of *60 Minutes* segment "Warning: May Be Fatal," box 1188, 77-3, *William P. Moore v. Allied Chemical Corporation*, file 6, "CA 77-0379-R, NARA-Philadelphia.

60. *Kepone Contamination*, 4.

61. "Governor's Press Release," December 17, 1975, box 8, "State Board of Health Emergency Rules—Closing of James River 1975" folder, Virginia Department of Health, Office of the State Health Commissioner, Regulatory and Case Files, 1975–1997, Library of Virginia, Richmond.

62. See "Wingspread States on the Precautionary Principle," in "Urban Governance," Global Development Research Center, http://www.gdrc.org/u-gov/precaution-3.html.

63. Kerry H. Whiteside and Robert Gottlieb, *Precautionary Politics: Principle and Practice in Confronting Environmental Risk* (Cambridge: MIT Press, 2006), viii.

64. Nancy Langston, "The Retreat from Precaution: Regulating Diethylstilbestrol (DES), Endocrine Disruptors, and Environmental Health," *Environmental History* 13, no. 1 (2008): 42.

65. "Emergency Rule, State Board of Health, Closing of the James River and Its Tributaries to the Taking of Fish," December 18, 1975, box 8, "State Board of Health Emergency Rules—Closing of James River 1975" folder, Virginia Department of Health, Office of the State Health Commissioner, Regulatory and Case Files, 1975–1997, Library of Virginia, Richmond.

66. Jerry Lazarus, "Fishermen Criticize James Ban," *Richmond Times-Dispatch*, December 18, 1975.

67. "Emergency Rule, State Board of Health, Closing of the James River and its Tributaries to the Taking of Fish," December 18, 1975, box 8, "State Board of Health Emergency Rules: Closing of the James River, 1975" folder, Virginia Department of Health, Office of the State Health Commissioner, Regulatory and Case Files, 1975–1997, Library of Virginia.

68. "Kepone Takes Toll on Family," *Harrisonburg Daily News-Record*, December 24, 1975.

69. Neil Cotiaux, interview with the author, October 19, 2016.

Chapter 4. Biocitizenship and Accountability

1. Vanessa Agard-Jones, "Bodies in the System," *Small Axe* 17, no. 3 (2013): 187.

2. Christopher Sellers, *Hazards of the Job* (Chapel Hill: University of North Carolina Press, 1997), 8.

3. Robert N. Proctor and Londa Schiebinger, eds., *Agnotology: The Making and Unmaking of Ignorance* (Stanford: Stanford University Press, 2008), vii.

4. "What they know, rather than what they do not know": Scott Frickel, "On Missing New Orleans: Lost Knowledge and Knowledge Gaps in an Urban Hazardscape," *Environmental History* 13 (2008): 645.

5. The most famous work uncovering these issues is Naomi Oreskes and Erik M. Conway, *Merchants of Doubt: How a Handful of Scientists Obscured the Truth on Issues from Tobacco Smoke to Climate Change* (New York: Bloomsbury, 2011).

6. David Weir and Mark Schapiro, *Circle of Poison: Pesticides and People in a Hungry World* (San Francisco: Institute for Food and Development Policy, 1981).

7. Kelly E. Happe, Jenell Johnson, and Marina Levina, eds., *Biocitizenship: The Politics of Bodies, Governance, and Power* (New York: New York University Press, 2018); Adriana Petryna, *Life Exposed: Biological Citizens after Chernobyl* (Princeton: Princeton University Press, 2002); and Nikolas Rose, *The Politics of Life Itself: Biomedicine, Power, and Subjectivity in the Twenty-First Century* (Princeton: Princeton University Press, 2007).

8. On how whiteness in recent years often puts lower-income workers at risk, see Jonathan Metzl, *Dying of Whiteness: How the Politics of Racial Resentment Is Killing America's Heartland* (New York: Basic Books, 2019).

9. Michelle Brattain, *The Politics of Whiteness: Race, Workers, and Culture in the Modern South* (Athens: University of Georgia Press, 2004), 6.

10. Brattain, *Politics of Whiteness*, 36.

11. Elizabeth D. Blum, *Love Canal Revisited: Race, Class, and Gender in Environmental Activism* (Lawrence: University Press of Kansas, 2008); Robert D. Bullard, *Dumping in Dixie: Race, Class, and Environmental Quality*, 3rd ed. (Boulder, Colo.: Routledge, 2000); Steve Lerner and Phil Brown, *Sacrifice Zones: The Front Lines of Toxic Chemical Exposure in the United States* (Cambridge, Mass.: MIT Press, 2012); Steve Lerner and Robert D. Bullard, *Diamond: A Struggle for Environmental Justice in Louisiana's Chemical Corridor* (Cambridge, Mass.: MIT Press, 2006); Richard S. Newman, *Love Canal: A Toxic History from Colonial Times to the Present* (New York: Oxford University Press, 2019); Bryant Simon, *The Hamlet Fire: A Story of Cheap Food, Cheap Government, and Cheap Lives* (New York: New Press, 2017); and Ellen Griffith Spears, *Baptized in PCBs: Race, Pollution, and Justice in an All-American Town*, New Directions in Southern Studies (Chapel Hill: University of North Carolina Press, 2014).

12. *Kepone Contamination: Hearings before the Subcommittee on Agricultural Research and General Legislation of the Committee on Agriculture and Forestry*, U.S. Senate, 94th Cong., 2nd sess., 131 (1976), http://hdl.handle.net/2027/mdp.39015071090206.

13. *Oversight Hearings on the Occupational Safety and Health Act: Hearings before the Subcommittee on Manpower, Compensation, and Health and Safety of the Committee on Education and Labor, House of Representatives*, 94th Cong., 2nd sess. (1976), 28–29.

14. *Oversight Hearings on the Occupational Safety and Health Act*, 2.

15. Donald (Tom) Fitzgerald, interview with the author, December 5, 2016.

16. *Oversight Hearings on the Occupational Safety and Health Act*, 37.

17. "10 File $25 Million Suit against 2 Firms on Kepone," *Richmond Times-Dispatch*, September 20, 1975.

18. "$24 Million Suit Filed against Allied, Hooker," *Hopewell News*, September 22, 1975.

19. Nicholas Brown and Dale Eisman, "Kepone Suits Increase in 2 Courts," *Richmond Times-Dispatch*, February 25, 1976; Dale Eisman and Nicholas Brown, "Three Kepone Suits Are Settled Here," *Richmond Times-Dispatch*, December 31, 1976.

20. "Complaint," *Gilbert v. Allied*, September 19, 1975, box 1057, case 75-3, *Gilbert v. Allied* file 1, NARA-Philadelphia.

21. "Request for Admission," *Gilbert v. Allied*, December 29, 1975, box 1058, case 75-3, *Gilbert v. Allied* file 3, NARA-Philadelphia.

22. "Answers and Objections to Plaintiffs' Request for Admission," *Gilbert v. Allied*, December 29, 1975, box 1058, case 75-3, *Gilbert v. Allied* file 3, NARA-Philadelphia.

23. "Answers and Objections to Plaintiffs' Request for Admission," *Gilbert v. Allied*, December 29, 1975, box 1058, case 75-3, *Gilbert v. Allied* file 3, NARA-Philadelphia.

24. "Allied Denies Kepone Illness to Blame," *Washington Star*, December 27, 1975.

25. "Allied Held to Have Known Kepone was Toxic," *New York Times*, December 25, 1975.

26. Sam Roberts, "Dr. Samuel Epstein, 91, Cassandra of Cancer Prevention, Dies," *New York Times*, April 25, 2018.

27. Samuel S. Epstein, "Preliminary Report on the Experimental Toxicology of Kepone," June 14, 1976, *Gilbert v. Allied*, December 29, 1975, box 1058, file 3, NARA-Philadelphia.

28. Rudolph J. Jaeger, "Final Report," June 18, 1976, *Gilbert v. Allied*, December 29, 1975, box 1058, 75-3, *Gilbert v. Allied*, file 3, NARA-Philadelphia.

29. Rudolph J. Jaeger, "Kepone Chronology," letter, *Science*, July 9, 1976, 94.

30. *Kepone Contamination*, 206–7.

31. *Kepone Contamination*, 163, 201.

32. *Kepone Contamination*, 172, 192.

33. *Oversight Hearings on the Occupational Safety and Health Act*, 115, 127.

34. *Kepone Contamination*, 196.

35. "Plaintiff's Brief in Reply to Briefs of Allied Chemical Corporation, Hooker Chemicals, and The Travelers Indemnity Company," 1979, page 6, box 3, "*William P. Moore v. Allied Chemical*" folder, C. Brian Kelly Papers, Albert and Shirley Small Special Collections Library, University of Virginia, Charlottesville.

36. "Plaintiff's Brief in Reply to Briefs of Allied Chemical Corporation," 10.

37. "Plaintiff's Brief in Reply to Briefs of Allied Chemical Corporation," 11.

38. *Oversight Hearings on the Occupational Safety and Health Act*, 114–15.

39. *Oversight Hearings on the Occupational Safety and Health Act*, 127.

40. *Oversight Hearings on the Occupational Safety and Health Act*, 114, 130.

41. "Plaintiff's Brief in Reply to Briefs of Allied Chemical Corporation."

42. *Kepone Contamination*, 257, 263.

43. *Kepone Contamination*, 277.

44. *Kepone Contamination*, 263.

45. *Kepone Contamination*, 273.

46. Janice Denton, interview with the author, December 2, 2016.

47. *Kepone Contamination*, 176.

48. *Kepone Contamination*, 177, 189.

49. *Kepone Contamination*, 427.

50. *Oversight Hearings on the Occupational Safety and Health Act*, 127.

51. *Kepone Contamination*, 191.
52. *Oversight Hearings on the Occupational Safety and Health Act*, 114.
53. *Oversight Hearings on the Occupational Safety and Health Act*, 121.
54. *Oversight Hearings on the Occupational Safety and Health Act*, 116–17.
55. *Kepone Contamination*, 259, 287.
56. *Kepone Contamination*, 261–62, 269.
57. *Kepone Contamination*, 273.
58. *Kepone Contamination*, 276.
59. *Kepone Contamination*, 288.
60. *Kepone Contamination*, 266.
61. *Kepone Contamination*, 156.
62. Frank Arrigo, interview with the author, March 18, 2017.
63. Steve Keavy, interview with the author, December 5, 2016.
64. *Kepone Contamination*, 139, 145.
65. *Kepone Contamination*, 323–26.
66. *Oversight Hearings on the Occupational Safety and Health Act*, 30–32.
67. Fitzgerald, interview.
68. *Oversight Hearings on the Occupational Safety and Health Act*, 41–44.
69. *Oversight Hearings on the Occupational Safety and Health Act*, 37.
70. Fitzgerald, interview.
71. *Oversight Hearings on the Occupational Safety and Health Act*, 26.
72. *Kepone Contamination*, 125–27.
73. *Oversight Hearings on the Occupational Safety and Health Act*, 20.
74. *Oversight Hearings on the Occupational Safety and Health Act*, 28–29.
75. *Oversight Hearings on the Occupational Safety and Health Act*, 21.
76. John Taylor, interview with the author, October 31, 2017.
77. Richard Foster, "Kepone: The 'Flour' Factory," *Richmond Magazine*, July 8, 2005, https://richmondmagazine.com/news/kepone-disaster-pesticide.
78. Jim Roberts, "One Man's Poison Is Another's Drink," *Richmond Times-Dispatch*, June 11, 1976.
79. Ronald J. Bacigal and Margaret I. Bacigal, "Criminal Prosecutions in Environmental Law: A Study of the 'Kepone' Case," *Columbia Journal of Environmental Law* 12, no. 291 (1987): 299.
80. "Drug That Speeds Kepone from Body Reported Found," *New York Times*, February 2, 1978.
81. William J. Cohn et al., "Treatment of Chlordecone (Kepone) Toxicity with Cholestyramine," *New England Journal of Medicine* 293, no. 5 (1978): 243–48.
82. Philip Guzelian et al., "Liver Structure and Function in Patients Poisoned with Chlordecone (Kepone)," *Gastroenterology* 78, no. 2 (1980): 206–13. See also "Kepone Rechecks Due," *Hopewell News*, July 18, 1980.
83. *Toxicological Review of Chlordecone (Kepone)*, CAS 143-50-0 (Environmental Protection Agency, 2009), 99, https://cfpub.epa.gov/ncea/iris/iris_documents/documents/toxreviews/1017tr.pdf.
84. "Kepone Rechecks Due," *Hopewell News*, July 18, 1980, 2.
85. Philip Guzelian, "Comparative Toxicology of Chlordecone (Kepone) in Humans and Experimental Animals," *Annual Review of Pharmacology and Toxicology* 22 (1982): 90.
86. "Kepone Contamination Effects Still Linger," *Winchester (Va.) Star*, July 11, 1985.

87. Richard E. Gordon, "Kepone: Chemical Remains Fact of Life for Its Makers and Watermen," *Richmond Times-Dispatch*, June 9, 1985.

88. Bernard Crutzen, *Pour quelques bananes de plus, le scandale du chlordécone* (Camera One Television, 2019).

89. Robert H. Schmerling, "Scleroderma," Harvard Health Publishing, October 20, 2020, https://www.health.harvard.edu/a_to_z/scleroderma-a-to-z.

90. Crutzen, *Pour quelques bananes de plus*.

91. *Oversight Hearings on the Occupational Safety and Health Act*, 22.

92. William Cummings, interview with the author, March 25, 2020.

93. Patricia Sullivan, "Federal Judge Robert J. Merhige Dies," *Washington Post*, February 20, 2005; "School Busing," Virginia Museum of History and Culture, accessed May 24, 2022, https://virginiahistory.org/learn/historical-book/chapter/school-busing; and Ron Wolf, "A Judge Unafraid of Unpopular Rulings," *Philadelphia Inquirer*, December 12, 1987.

94. Mark Merhige, interview with the author, October 17, 2016.

95. Bacigal and Bacigal, "Criminal Prosecutions in Environmental Law: A Study of the 'Kepone' Case"; Robert R. Merhige Jr. et al., "Panel Discussions: Allied Chemical, the Kepone Incident, and the Settlements: Twenty Years Later," *University of Richmond Law Review* 29, no. 493 (1995); and Eston Melton, "Merhige Delays Duke Tests, Hints at Kepone Trial Delay," *Richmond Times-Dispatch*, August 9, 1976.

96. Merhige et al., *Panel Discussions*, 514–15.

97. Marvin Zim, "Allied Chemical's $20-Million Ordeal with Kepone," *Fortune*, September 11, 1978, 82–84, 88–90.

98. Eisman and Brown, "Three Kepone Suits Are Settled Here."

99. "Order," December 22, 1980, box 1212, 77-3, *Donald F. Case, Jr. and Carl K. Miles, Jr. v. Allied Chemical Corporation*, CA 77-0187-R and 77-0188-R, file no. 6, NARA-Philadelphia.

100. Merhige et al., "Panel Discussions," 496.

101. Transcript, Crutzen, *Pour quelques bananes de plus*. In 2019 terms, $64,000 is about $287,000.

102. Crutzen, *Pour quelques bananes de plus*.

103. Richard E. Gordon, "Kepone: Chemical Remains Fact of Life for Its Makers and Watermen," *Richmond Times-Dispatch*, June 9, 1985.

104. Janice Denton, interview with the author, December 2, 2016.

105. Steve Keavy, interview with the author, December 5, 2016.

Chapter 5. A Crime against Every Citizen

1. Kit Weigel, "Seal Failure Sets Opening of New Plant," *Hopewell News*, February 22, 1974.

2. Memorandum to File, March 28, 1974, Jon R. Carroll, Virginia Air Pollution Control Board, "Life Science Products Company," box 1212, 77-3, *Donald F. Case, Jr., v. Allied Chemical Corporation, et al.*, CA 77-0188-R, file 5, Virginia–Eastern District, Records of District Courts of the United States, RG 21, NARA-Philadelphia; and "Sulfur Trioxide (SO_3) and Sulfuric Acid," Agency for Toxic Substances and Disease Registry, June 1999, https://www.atsdr.cdc.gov/toxfaqs/tfacts117.pdf.

3. "Seal Failure Sets Opening of New Plant," *Hopewell News*, February 22, 1974.

4. Memorandum to File, March 28, 1974.

5. Nicholas Brown, "Plant Wasn't Inspected," *Richmond Times-Dispatch*, January 18, 1976.

6. "Chronology of Events Relevant to Production of Kepone by Life Science Products, Inc.,

Hopewell, Virginia (1974–1975)," James B. Kenley, M.D., Deputy Commissioner of Health, to Otis Brown, Secretary of Human Affairs, and Earl J. Shiflet, Secretary of Commerce and Resources, February 19, 1976, box 8, folder 1, Herbert H. Bateman State Senate Papers, Special Collections, Swem Library, College of William & Mary, Williamsburg, Virginia.

7. Memorandum to File, March 28, 1974.

8. Minutes: June 12, 1972, box 1, "Board Meeting: 6/17/72," folder, Department of Air Pollution Control, Board Meeting Records, 5-17-72 to 5-6-74, LVA.

9. *Kepone Contamination: Hearings before the Subcommittee on Agricultural Research and General Legislation of the Committee on Agriculture and Forestry, U.S. Senate*, 94th Cong., 2nd Sess., 77 (1976), http://hdl.handle.net/2027/mdp.39015071090206.

10. *Oversight Hearings on the Occupational Safety and Health Act: Hearings before the Subcommittee on Manpower, Compensation, and Health and Safety of the Committee on Education and Labor, House of Representatives*, 94th Cong., 2nd sess. (1976), 135–46.

11. *Kepone Contamination*, 356.

12. William H. Rodgers Jr., "Industrial Water Pollution and the Refuse Act: A Second Chance for Water Quality," *University of Pennsylvania Law Review* 119 (1971): 761–822.

13. See "Water Permitting 101," Office of Wastewater Management, U.S. Environmental Protection Agency, accessed June 19, 2019, https://www3.epa.gov/npdes/pubs/101pape.pdf.

14. *Kepone Contamination*, 31–37.

15. James Ezzell, "Hopewell Was Moving on Kepone Firm," *Richmond Times-Dispatch*, January 19, 1976.

16. Memorandum, "Life Science Products Company et al.," Andrew Miller, Attorney General, to State Water Control Board, January 20, 1976, box 9, "Kepone" folder, General Correspondence: Gunston Hall-Kepone, Executive Papers of Governor Mills E. Godwin Jr., Library of Virginia, Richmond; James Ezzell, "Hopewell Was Moving on Kepone Firm," *Richmond Times-Dispatch*, January 19, 1976.

17. P. Aarne Vesilind and Alastair S. Gunn, *Engineering, Ethics, and the Environment* (New York: Cambridge University Press, 1998), 146.

18. *Kepone Contamination*, 53, 169.

19. James Ezzell, "Hopewell Finds No Record of Filing EPA Amendment," *Richmond Times-Dispatch*, January 21, 1976.

20. Kit Weigel, "Havens Tends to City 'Stomach,'" *Hopewell News*, May 1, 1975; and Vesilind and Gunn, *Engineering, Ethics, and the Environment*, 146.

21. Kit Weigel, "Bailey's Creek Has Flu; Suffers from Lack of 'Bugs,'" *Hopewell News*, November 7, 1974; "Bugless Sludge Put in Special Pit," *Hopewell News*, May 12, 1975.

22. Weigel, "Bailey's Creek Has Flu"; and Dale Eisman, "Allied Trial Offered Scenario of Kepone Pollution," *Richmond Times-Dispatch*, October 4, 1976.

23. Memorandum, "Life Science Products Company et al."

24. Dale Eisman, "Kepone Firm and Allied Independent, Moore Says," *Richmond Times-Dispatch*, February 28, 1976.

25. Memorandum, "Life Science Products Company, et al.—Violations of the State Water Control Law," James E. Kulp to Mills Godwin, Jr., February 12, 1976, box 9, "Kepone" folder, RG 3, Executive Papers of Governor Mills E. Godwin Jr., Library of Virginia, Richmond.

26. "Chronology of Events Relevant to Production of Kepone."

27. *Kepone Contamination*, 105–6.

28. *Kepone Contamination*, 101, 206, 241; Letter, William P. Moore to I. Swisher, 10 February 1975, box 1188,77-3, *William P. Moore v. Allied Chemical*, file 6, NARA-Philadelphia.

29. *Kepone Contamination*, 101.

30. Memorandum, J. B. Reeves to J. J. Cibulka, November 12, 1974, in *Kepone Contamination*, 100–01.

31. Weigel, "Bailey's Creek Has Flu," 12.

32. James Ezzell, "Hopewell Was Moving on Kepone Firm," *Richmond Times Dispatch*, January 19, 1976.

33. *Kepone Contamination*, 99.

34. *Kepone Contamination*, 99.

35. *Kepone Contamination*, 107.

36. *Kepone Contamination*, 170.

37. *Kepone Contamination*, 82.

38. *Kepone Contamination*, 54.

39. Agency for Toxic Substances and Disease Registry, *Mirex and Chlordecone* (Washington, D.C.: Agency for Toxic Substances and Disease Registry, 1995); and World Health Organization, *Mirex*, Environmental Health Criteria 44 (Geneva, Switzerland: World Health Organization, 1984).

40. *Kepone Contamination*, 38, 53.

41. *Kepone Contamination*, 92.

42. Steve Keavy, interview with the author, December 5, 2016.

43. Memorandum, "Life Science Products Company et al.," Andrew Miller.

44. Virginia General Assembly, Joint Legislative Audit and Review Commission, *Program Evaluation: Water Resource Management in Virginia* (Richmond: The Commission, 1976).

45. James Ryan, interview with the author, December 13, 2016.

46. Memorandum, "Life Science Products Company, et al.—Violations of the State Water Control Law," James E. Kulp to Mills Godwin Jr.

47. *Kepone Contamination*, 37–38, 94–98.

48. *Kepone Contamination*, 309.

49. *Kepone Contamination*, 38–41.

50. *Oversight Hearings on the Occupational Safety and Health Act*, 233.

51. Nicholas Brown, "Worker in Middle as 'Watchdogs' Feud," *Richmond Times-Dispatch*, September 28, 1975.

52. Nicholas Brown, "Plant Wasn't Inspected," *Richmond Times-Dispatch*, January 18, 1976.

53. Robert R. Merhige Jr. et al., "Panel Discussions: Allied Chemical, the Kepone Incident, and the Settlements: Twenty Years Later," *University of Richmond Law Review* 29, no. 493 (1995): 511.

54. "Statement of Mr. John T. Connor Concerning Settlement of Kepone Claims with Commonwealth of Virginia and City of Hopewell," C. Brian Kelly Papers, box 2, "Kepone--Allied Chemical" folder, Albert and Shirley Small Special Collections Library, University of Virginia, Charlottesville.

55. Marvin Zim, "Allied Chemical's $20-Million Ordeal with Kepone," *Fortune*, September 11, 1978, 82–84, 88–91.

56. *Kepone Contamination*, 204–9.

57. William Cummings, interview with the author, March 25, 2020.

58. Quoted in Brian Kelly, "U.S. May Ask Jury to Study Kepone Case," *Washington Star*, December 13, 1975.

59. Merhige et al., "Panel Discussions," 504.

60. Merhige et al., "Panel Discussions," 504.

61. Richard Foster, "Kepone: The 'Flour' Factory," *Richmond Magazine*, July 8, 2005, https://richmondmagazine.com/news/kepone-disaster-pesticide.

62. James Ezzell, "'74 Air Tests Found Allied-Made Kepone," *Richmond Times-Dispatch*, January 27, 1976; and Jeff Douglas, "Kepone: An Attack on Fabric of Man and Nature," *Radford (Va.) News Journal*, February 3, 1976. Hargis faced controversy over his management of VIMS, coming under fire from Godwin for his budgetary issues, but also his outspoken views on Godwin's projects and environmental issues facing Virginia's waterways and the Chesapeake Bay. Amid the Kepone crisis, the state police launched an investigation against him. He had charges of theft dismissed, but he lost his control of VIMS in 1981; the agency also moved under the control of the College of William & Mary.

63. Merhige et al., "Panel Discussions," 507–8.

64. Diane D. Eames, "The Refuse Act of 1899: Its Scope and Role in Control of Water Pollution," *California Law Review* 58, no. 6 (1970): 1450.

65. Zygmunt J. B. Plater, *Environmental Law and Policy: Nature, Law and Society*, 4th ed. (New York: Aspen, 2010), 40–41. See also "ACC Men Lied, Says Huntdofte at Trial," *Hopewell News*, September 1, 1976.

66. Zim, "Allied Chemical's $20-Million Ordeal."

67. James Ezzell, "Kepone Waste Was Dumped into James for Seven Years," *Richmond Times-Dispatch*, January 22, 1976.

68. Cummings, interview.

69. William Goldfarb, "Kepone: A Case Study," *Environmental Law* 8, no. 3 (1978): 648; and "Kepone Charges in Detail," *Washington Post*, May 8, 1976.

70. Kit Weigel, "City Said Guilty in Kepone Trial," *Hopewell News*, June 25, 1976, 1.

71. Merhige et al., "Panel Discussions," 496.

72. Goldfarb, "Kepone," 655.

73. Emily England, "Allied Cleared of Felony Count," *Progress-Index* (Petersburg, Va.), July 8, 1976.

74. "Plea Bargain Said Key for Other Kepone Cases," *Hopewell News*, August 11, 1976; "Kepone Company Pleads Not Guilty," *Richmond Times Dispatch*, August 17, 1976.

75. Emily England, "Allied Will Not Contest Charges of River Pollution," *Progress-Index* (Petersburg, Va.), August 20, 1976; and Ronald J. Bacigal and Margaret I. Bacigal, "Criminal Prosecutions in Environmental Law: A Study of the 'Kepone' Case," *Columbia Journal of Environmental Law* 12, no. 291 (1987): 300.

76. "Allied Admits Pollution Guilt in Own Operation," *Hopewell News*, August 20, 1976.; and Bacigal and Bacigal, "Criminal Prosecutions in Environmental Law," 301–2.

77. Ben Franklin, "Two Executives Acquitted of Plotting in Kepone Case," *New York Times*, September 2, 1976.

78. Merhige et al., "Panel Discussions," 508.

79. Emily England, "Moore Pleads No Contest to Misdemeanor Pollution Counts," *Progress-Index* (Petersburg, Va.), September 15, 1976.

80. Bacigal and Bacigal, "Criminal Prosecutions in Environmental Law," 301.

81. Goldfarb, "Kepone," 658.

82. Emily England, "Huntdofte Tells of Meeting," *Progress-Index* (Petersburg, Va.), September 28, 1976.

83. Letter, Moore to Swisher, March 14, 1975, box 1188, 77-3, *William P. Moore v. Allied Chemical*, file 6, NARA, Philadelphia.

84. Bacigal and Bacigal, "Criminal Prosecutions in Environmental Law," 305.

85. Bacigal and Bacigal, "Criminal Prosecutions in Environmental Law," 305.

86. Brad Holt, "Court Told of Kepone Discharges," *Washington Star*, September 29, 1976.

87. Dale Eisman, "Allied Trial Offered Scenario of Kepone Pollution," *Richmond Times-Dispatch*, October 4, 1976; Kit Weigel, "Moore Called 'Wheeler-Dealer' in Kepone Trial," *Hopewell News*, September 29, 1976; Dale Eisman, "Hundtofte Contradicts Self on Shipments from Allied," *Richmond Times-Dispatch*, September 29, 1976; Bill McCallister, "2 Tell of Warning Corporation on Kepone Pollution," *Washington Post*, September 29, 1976.

88. Dale Eisman, "Allied Granted Acquittals on 144 Abetting Counts," *Richmond Times-Dispatch*, September 30, 1976; and Bacigal and Bacigal, "Criminal Prosecutions in Environmental Law," 304.

89. Bacigal and Bacigal, "Criminal Prosecutions in Environmental Law," 306.

90. Bacigal and Bacigal, "Criminal Prosecutions in Environmental Law," 307–08.

91. Bacigal and Bacigal, "Criminal Prosecutions in Environmental Law," 306–8; Dale Eisman, "Kepone Case Conspiracy Not Proved," *Richmond Times Dispatch*, October 1, 1976; Kit Weigel, "Two Witnesses Tilted Balance," *Hopewell News*, October 1, 1976; and Zim, "Allied Chemical's $20-Million Ordeal."

92. Merrill Brown, "EPA to Let Mississippi Produce Ant Killer," *Richmond Times-Dispatch*, May 9, 1976.

93. Bill Hoyle, "Kepone: Two Years Later, No End in Sight as Costs Mount," *Richmond Times-Dispatch*, July 24, 1977.

94. The phrase "crime against every citizen" does not appear in the transcript of Judge Merhige's ruling as read from the bench. He added this in making the ruling public, perhaps to reporters covering the trial. See "Ruling as Stated from the Bench," "Legislation/Registration" folder, Kepone records, Virginia Department of Environmental Quality, Richmond; Dale Eisman, "$13.24 Million Allied Fine Largest Ever for Pollution," *Richmond Times-Dispatch*, October 6, 1976; and Emily England, "$13 Million Allied Fine Is Record," *Progress-Index*, (Petersburg, Va.), October 6, 1976.

95. Bacigal and Bacigal, "Criminal Prosecutions in Environmental Law," 312.

96. Otis Brown, interview with the author, June 24, 2018. Trowbridge had succeeded Allied's John T. Connor as commerce secretary under Lyndon Johnson, and Connor recruited him to Allied.

97. Ray McAllister, "Allied Endowment May Reduce Fine," *Richmond Times-Dispatch*, January 29, 1977.

98. Joseph H. Maroon, "40 Years of Bettering Virginia's Environment," Virginia Environmental Endowment, http://www.vee.org/wp-content/uploads/2013/07/VEE-40th-Op-Ed-FINAL.pdf.

99. "Judge Praises Allied; Fine Cut to $5 million," *Hopewell News*, February 2, 1977.

100. Merhige et al., "Panel Discussions," 509.

101. Gerald P. McCarthy, oral history interview, Virginia Museum of History & Culture, Richmond; Virginia Environmental Endowment, Annual Report, 1977, in Virginia Environmental Endowment Records, Annual Reports, folder 93, Virginia Museum of History & Culture, Richmond.

102. "Judge Praises Allied; Fine Cut to $5 million," *Hopewell News*, February 2, 1977.

103. Merhige et al., "Panel Discussions," 509.

104. McCarthy, interview.

105. Bacigal and Bacigal, "Criminal Prosecutions in Environmental Law," 313.

106. Allied Chemical, "Allied Chemical Annual Report, 1977," February 24, 1978.

107. Merhige et al., "Panel Discussions," 513.

108. Brent Fisse and John Braithwaite, "Corporate Offences: The Kepone Affair," in *Combating Commercial Crime*, ed. Rae Watson(Sydney, Australia: Law Book, 1987), 31–51.

109. Quoted in Emily England, "Fishermen's Suit May Include City," *Progress-Index* (Petersburg, Va.), January 21, 1976.

110. Emily England, "Hopewell Joins Kepone Related Defendant List," *Progress-Index* (Petersburg, Va.), March 25, 1976.

111. "$500,000 Said Paid for Kepone Damages," *Hopewell News*, June 26, 1980.

112. "Pruitt v. Allied Chemical Corp., 85 F.R.D. 100 (1980)," Caselaw Access Project, accessed August 9, 2020, https://cite.case.law/frd/85/100.

113. "Pruitt v. Allied Chemical Corp., 85 F.R.D. 100 (1980)."

114. Andrew W. McThenia and Joseph E. Ulrich, "A Return to First Principles of Corrective Justice in Deciding Economic Loss Cases," *Virginia Law Review* 69, no. 8 (1983): 1518.

115. David Frisch, "It's About Time," *Tennessee Law Review* 79 (2012): 785.

116. "Pruitt v. Allied Chemical Corp., 523 F. Supp. 975 (E.D. Va. 1981)," Justia Law, accessed August 14, 2020, https://law.justia.com/cases/federal/district-courts/FSupp/523/975/2298321.

117. "Pruitt v. Allied Chemical Corp., 523 F. Supp. 975"; and Rob Walker, "Accord Settles Last Kepone Suit," *Richmond Times-Dispatch*, June 12, 1982.

118. "Kepone Waste Discharge Civil Suit Filed by State," *Progress-Index* (Petersburg, Va.), March 22, 1976.

119. James Ryan, interview with the author, December 13, 2016.

120. Bill Hoyle, "Kepone: Two Years Later, No End in Sight as Costs Mount," *Richmond Times-Dispatch*, July 24, 1977; and Ryan, interview.

121. Memo, James E. Ryan Jr., August 25, 1977, in box 51, "Kepone" folder, Executive Papers of Governor Mills E. Godwin Jr., Library of Virginia, Richmond.

122. Otis Brown, interview with the author, June 24, 2018.

123. Bill Hoyle, "Allied to Pay State, Hopewell $5.2 Million," *Richmond Times-Dispatch*, October 14, 1977; "Mutual Release," box 51, "Kepone" folder, Letters Received and Sent, Executive Papers of Governor Mills E. Godwin Jr.

124. Michael R. Reich and Jaquelin K. Spong, "Kepone: A Chemical Disaster in Hopewell, Virginia," *International Journal of Health Services* 13, no. 2 (1983): 238.

125. Gabrielle Hecht, "Residue," *Somatosphere* (blog), January 8, 2018, http://somatosphere.net/2018/residue.html.

Chapter 6. Kepone and the Environmental Management State

1. "Fish Consumption Advisory," Virginia Department of Health, https://www.vdh.virginia.gov/environmental-health/public-health-toxicology/fish-consumption-advisory.

2. On the development of these trends, especially in the 1970s, see Naomi Oreskes and Erik M. Conway, *Merchants of Doubt: How a Handful of Scientists Obscured the Truth on Issues from Tobacco Smoke to Climate Change* (New York: Bloomsbury, 2011); and Robert N. Proctor, *Cancer Wars: How Politics Shapes What We Know and Don't Know about Cancer* (New York: Basic Books, 1995).

3. Andrew C. Isenberg, ed., *The Oxford Handbook of Environmental History* (Oxford, UK: Oxford University Press, 2014), 729.

4. Lee Clarke, *Acceptable Risk? Making Decisions in a Toxic Environment* (Berkeley: University of California Press, 1991); Kathryn Harrison and George Hoberg, *Risk, Science, and Politics: Regulating Toxic Substances in Canada and the United States* (Montreal: McGill-Queen's University Press, 1994); Nancy Langston, *Toxic Bodies: Hormone Disruptors and the Legacy of DES* (New Haven, Conn.: Yale University Press, 2010); and Sarah A. Vogel, *Is It Safe? BPA and the Struggle to Define the Safety of Chemicals* (Berkeley: University of California Press, 2013).

5. Stephen Bocking, *Nature's Experts: Science, Politics, and the Environment* (Piscataway, N.J.: Rutgers University Press, 2004), 137.

6. Bocking, *Nature's Experts*, 144–46.

7. Kepone Task Force, "Meeting Number 8: Highlights," January 16, 1976, box 8, folder 4, Herbert H. Bateman State Senate Papers, Special Collections, Swem Library, College of William & Mary, Williamsburg, Virginia; and Kepone Task Force Meeting, "Minutes," January 16, 1976, box 21, "Kepone Task Force Minutes and Agenda" folder, Council on the Environment, General Correspondence and Reference Files, 1971–1980, Library of Virginia, Richmond.

8. James E. Ryan, interview with the author, December 16, 2016.

9. James E. Ryan, email communication to author, November 23, 2020.

10. Gerald Baliles, interview with the author, November 15, 2016. Governor Baliles passed away in 2019.

11. Gerald Baliles, oral history interview, Virginia Museum of History and Culture, Richmond.

12. Gerald Baliles, interview with the author, November 15, 2016.

13. Ryan, email.

14. Nicholas Brown, "Kepone-Prompted Bills Seen as Costly but Effective Laws," *Richmond Times-Dispatch*, March 22, 1976.

15. Jerry McCarthy, handwritten notes, box 21, "Kepone Task Force, General Correspondence" folder, CEA General Correspondence and Reference Files, LVA.

16. *Worker Safety in Pesticide Production: Hearings before the Subcommittee on Agricultural Research and General Legislation of the Committee on Agriculture, Nutrition, and Forestry, U.S. Senate*, 95th Cong., 1st sess., 168 (1977).

17. "Toxic Substances Information Act," box 27, "H.B. 1454, Toxic Substances Information Act (1 of 2)," folder, Series 6: Subject Files, Gerald L. Baliles Papers, Virginia Museum of History and Culture, Richmond. See also *Acts of the General Assembly of the Commonwealth of Virginia, Session 1976, January 14, 1976, to March 13, 1976*, vol. 1 (Richmond: Commonwealth of Virginia, 1976), chap. 627, 852–55.

18. *Worker Safety in Pesticide Production*.

19. "Memorandum on Proposed Amendments to the Toxic Substances Information Act," box 27, "H.B. 1454, Toxic Substances Information Act (1 of 2)," folder, Series 6: Subject Files, Gerald L. Baliles Papers, Virginia Museum of History and Culture, Richmond.

20. "Toxin List Reduction Planned," *Daily Progress* (Charlottesville, Va.), November 18, 1976.

21. "Memorandum on Proposed Amendments to the Toxic Substances Information Act"; and "Statement of Gerald L. Baliles on Proposed Rules and Regulations Regarding Toxic Substances," July 14, 1976, Gerald L. Baliles Papers, Virginia Museum of History and Culture, Richmond.

22. *Report of the Joint Subcommittee Studying the Effectiveness of the Toxic Substances Information Act* (Richmond: Commonwealth of Virginia, Senate, 1982), 4.

23. Anne Hazard, "Toxic Chemical Data Virtually Ignored," *Richmond Times-Dispatch*, October 31, 1982.

24. *Report of the Joint Subcommittee Studying the Effectiveness of the Toxic Substances Information Act*, 3.

25. Joseph Gatins, "Damage Waiver Proposed for Giving Data in Error," *Richmond Times-Dispatch*, December 13, 1983; Bonnie V. Winston, "Confidential Data Legislation Will Aid State Employees," *Richmond Times-Dispatch*, March 30, 1984.

26. Michael Wines, "Carbide Lays Leak to Human, Machine Errors," *Los Angeles Times*, August 24, 1985.

27. Linda-Jo Schierow, *The Toxic Substances Control Act (TSCA): Implementation and New Challenges* (Washington, D.C.: Congressional Research Service, 2009); and Ray M. Druley and Girard Lanterman Ordway, *The Toxic Substances Control Act* (Washington, D.C.: Bureau of National Affairs, 1977), 11–12.

28. Schierow, *Toxic Substances Control Act*, 1.

29. Druley and Ordway, *Toxic Substances Control Act*, 16.

30. 94 Cong. Rec., S8296 (1976), https://www.govinfo.gov/content/pkg/GPO-CRECB-1976-pt7/pdf/GPO-CRECB-1976-pt7-3-1.pdf.

31. *Toxic Substances Control Act Hearings before the Subcommittee on Consumer Protection and Finance of the Committee on Interstate and Foreign Commerce, on H.R. 7229, H.R. 7548, and H.R. 7664, Bills to Regulate Commerce and Protect Human Health and the Environment by Requiring Testing and Necessary Use Restrictions on Certain Chemical Substances, and for Other Purposes, June 16, July 9, 10, and 11, 1975*, House of Representatives, 94th Cong., 1st sess., 1975.

32. Library of Congress, Environmental and Natural Resources Policy Division, *Legislative History of the Toxic Substances Control Act, Together with a Section-by-Section Index* (Washington, D.C.: U.S. Government Printing Office, 1976), 168, http://hdl.handle.net/2027/uc1.31210015688888.

33. Cong. Rec., 94th Cong., S8294 (1976).

34. Druley and Ordway, *Toxic Substances Control Act*, 23.

35. J. Brooks Flippen, *Conservative Conservationist: Russell E. Train and the Emergence of American Environmentalism* (Baton Rouge: Louisiana State University Press, 2006), 174. See also E. W. Kenworthy, "Train Asks Stiff Toxic Chemical Law," *New York Times*, February 27, 1976.

36. "Emergency Rule: Closing of the James River and Its Tributaries to the Taking of Fish," in box 8, "State Board of Health Emergency Rules–Closing of James River 1975" folder, Virginia Department of Health, Regulatory Case Files, 1975–1997, Library of Virginia, Richmond; "Kepone Report Coming," *Smithfield Times*, January 21, 1976.

37. "Emergency Rule"; Emily England, "Governor Relaxes Ban on James River Fishing," *Progress-Index* (Petersburg, Va.) February 3, 1976.

38. "Summary of Revisions of the Emergency Rule of the Virginia State Board of Health as Revised, April 14, 1976," box 8, "State Board of Health Emergency Rules—Closing of James River 1975" folder, Virginia Department of Health, Regulatory Case Files, 1975–1997, Library of Virginia, Richmond.

39. Linda Nash, "Purity and Danger: Historical Reflections on The Regulation of Environmental Pollutants," *Environmental History* 13 (October 2008): 654.

40. Thomas Truitt to Bateman, September 15, 1976, box 8, folder 21, and EPA "Fact Sheet: Action Levels for Pesticides," box 8, folder 22, Bateman Papers.

41. Thomas Truitt to Bateman; EPA Memo, January 5, 1976, box 8, folder 22, Bateman Papers.

42. This phrasing comes from Robert Proctor, author of *The Cancer Wars*.

43. *Establishment and Reviews of Original Kepone Action Levels* (Washington, D.C.: Environmental Protection Agency, 1977), 13.

44. "70 Scientists Agree Kepone No. 1 Enemy," *Progress-Index* (Petersburg, Va.), October 14, 1976.

45. "Emergency Rule."

46. "Ban James Fishing till '80, Panel Asks," *Richmond Times-Dispatch*, December 1, 1977.

47. John Dalton to James Kenley, August 4, 1978, and press release, December 1978, VDH, both in box 8, "State Board of Health Emergency Rules—Closing of James River 1975" folder, Virginia Department of Health, Regulatory Case Files, 1975–1997, Library of Virginia, Richmond.

48. Otis Brown, interview with the author, June 24, 2018.

49. Michael W. Fincham, "The Man Who Said Too Much: Bill Hargis and the Rise of a Marine Lab," *Chesapeake Quarterly*, April 2014, https://www.chesapeakequarterly.net/V13N1/main2.

50. James Kirkley, *Virginia's Commercial Fishing Industry: Its Economic Performance and Contributions* (Virginia Institute of Marine Science, 1997), http://publish.wm.edu/reports/361.

51. Dexter S Haven and James P. Whitcomb, "The Origin and Extent of Oyster Reefs in the James River, Virginia," *Journal of Shellfish Research* 3, no. 2 (1983): 141–51.

52. Kirkley, *Virginia's Commercial Fishing Industry*, 46.

53. Kirkley, *Virginia's Commercial Fishing Industry*, 47.

54. "Ban on Bluefish Not Yet Justified," *Progress-Index* (Petersburg, Va.), July 7, 1976; and "Kepone Sediment," *Winchester (Va.) Evening Star*, September 9, 1976.

55. *Kepone Contamination: Hearings before the Subcommittee on Agricultural Research and General Legislation of the Committee on Agriculture and Forestry, U.S. Senate*, 94th Cong., 2nd Sess., 245–47 (1976), http://hdl.handle.net/2027/mdp.39015071090206.

56. Morris Rowe, "Kepone Delivers More Punches to Crabbers," *Virginian-Pilot* (Norfolk), November 20, 1978.

57. David Ennis, "Crabbers Warned to Obey Laws," *Times-Herald* (Newport News, Va.), November 4, 1978.

58. James Kenley to John R. Broadway, Registrar of Regulations, June 21, 1979, Virginia Health Department, Office of the State Health Commissioner, Regulatory Case Files, 1975–1997, box 8, "State Board of Health Emergency Rules-Closing of the James River" folder, Library of Virginia, Richmond.

59. Michael Paolisso, "Blue Crabs and Controversy on the Chesapeake Bay: A Cultural Model for Understanding Watermen's Reasoning about Blue Crab Management," *Human Organization* 61, no. 3 (2002) 233. See also Christine Keiner, *The Oyster Question: Scientists, Watermen, and the Maryland Chesapeake Bay since 1880* (Athens: University of Georgia Press, 2009).

60. Keiner, *Oyster Question*, 15.

61. *Kepone Contamination*, 248–49.

62. Matthew Paust, "Kepone Tally Angers Seafood Industry," *Daily Press* (Newport News, Va.), July 3, 1976.

63. Gene Phillips, "VIMS Chief Takes Leave," *Daily Press* (Newport News, Va.), July 11, 1976.

64. Howard R. Ernst, *Chesapeake Bay Blues: Science, Politics, and the Struggle to Save the Bay* (Lanham, Md.: Rowman & Littlefield, 2003), 109.

65. The story he referenced was Soren Jensen's discovery of what became known as PCBs. Douglas did not quite get the metaphor correct, as DDT was present, and Jensen found another toxic substance. Soren Jensen, "The PCB Story," *Ambio* 1, no. 4 (1972): 123–31.

66. James E. Douglas to James Kenley, August 11, 1976, box 8, folder 1, Herbert H. Bateman State Senate Papers, Special Collections, Swem Library, College of William & Mary, Williamsburg, Virginia.

67. "Kepone Fact Analysis," box 8, folder 5, Bateman Papers.

68. Bateman to Godwin, August 24, 1976, box 8, folder 10, Bateman Papers.

69. Bateman to Jack Miles, President, Virginia Seafood Council, August 25, 1976, box 8, folder 10, Bateman Papers.

70. Among others, see Naomi Oreskes and Erik M. Conway, *Merchants of Doubt: How a Handful of Scientists Obscured the Truth on Issues from Tobacco Smoke to Climate Change* (New York: Bloomsbury, 2011); Robert N. Proctor, *Cancer Wars: How Politics Shapes What We Know and Don't Know About Cancer* (New York: Basic Books, 1995); and Nancy Langston, *Toxic Bodies: Hormone Disruptors and the Legacy of DES* (New Haven, Conn.: Yale University Press, 2010).

71. Proctor, *Cancer Wars*, 125.

72. Deichmann, "Curriculum Vitae," box 8, folder 8, Bateman Papers; and "Obituaries of Members of the Ohio Academy of Science: Report of the Necrology Committee, 1991," Knowledge Bank, Ohio State University, https://kb.osu.edu/bitstream/handle/1811/23480/V091N5_221.pdf.

73. William Bernard Deichmann, Sylvan Witherup, and Karl Kitzmiller, *The Toxicity of DDT: Experimental Observations* (Cincinnati: Kettering Laboratory in the Dept. of Preventive Medicine and Industrial Health, College of Medicine, University of Cincinnati, 1950).

74. Deichmann, "Cummings Memorial Lecture—1975: The Market Basket: Food for Thought," in Bateman Papers, box 8, folder 8. On the health and politics surrounding DDT, see David Kinkela, *DDT and the American Century: Global Health, Environmental Politics, and the Pesticide That Changed the World* (Chapel Hill: University of North Carolina Press, 2011).

75. Letter, Deichmann to Bateman, October 12, 1976, box 8, folder 8, Bateman Papers.

76. Langston, *Toxic Bodies*, 113.

77. William B. Deichmann, "Kepone: Review of the Acute and Chronic Toxicity and Action Level with Particular Reference to Fin Fish," November 10, 1976, box 8, folder 8, Bateman Papers.

78. Thomas Truitt to Bateman, February 3, 1977, box 8, folder 5, Bateman Papers. See also "Thomas Truitt Dies: Was Attorney in Love Canal Litigation," *Washington Post*, January 6, 2012.

79. Lee Weddig, National Fisheries Institute, to Jack Blanchard, EPA, October 14, 1976, box 8, folder 5, Bateman Papers.

80. "EPA Conference," handwritten notes, box 8, folder 14, Bateman Papers.

81. "Presentation by Herbert H. Bateman on Behalf of Virginia Seafood Industry Regarding Kepone Action Level, EPA Public Hearing, Richmond, Virginia, January 26, 1977," box 8, folder 16, Bateman Papers.

82. Statement of Dr. James Kenley, EPA Office of Pesticide Programs public hearing, January 26, 1977, box 8, folder 17, Bateman Papers.

83. Merrill Brown, "Alliance Asks Kepone Shifts," *Richmond Times-Dispatch*, January 27, 1977.

84. Statement of Michael E. Bender, PhD, EPA Office of Pesticide Programs public hearing, January 26, 1977, box 8, folder 17, Bateman Papers.

85. *Worker Safety in Pesticide Production: Hearings before the Subcommittee on Agriculture and General Legislation of the Committee on Agriculture, Nutrition, and Forestry*, U.S. Senate, 95th Cong., 1st sess. (1977), 104, 283, 285–87.

86. Robert Stroube, assistant state health commissioner, "Kepone Report," box 10, "Regulations Prohibiting the Taking of Crabs & Finfish From Chesapeake/James River" folder, Virginia Department of Health, Regulatory Case Files, 1975–1997, Library of Virginia, Richmond.

87. Richard Stradling, "Kepone Reshaped Life on the James," *Daily Press* (Newport News, Va.), July 23, 1995.

88. Cranston Morgan to Dalton, May 19, 1980, box 8, "State Board of Health Emergency Rules—Closing of James River 1975" folder, Virginia Department of Health, Regulatory Case Files, 1975–1997, Library of Virginia, Richmond.

89. Statement of Herbert Bateman, Public Hearing Regarding Ban of Fishing and Crabbing on the James River, July 2, 1980, box 8, folder 17, Bateman Papers.

90. Ron Sauder, "Fishing Ban, Kepone Levels Hit," July 3, 1980, *Richmond Times-Dispatch*; and "Public Hearing: Proposed Regulations to Ban Fishing and Crabbing on the James River and its Tributaries," July 2, 1980, box 10, "Regulations Prohibiting Taking of Crabs and Finfish From the Chesapeake/James River, 1980" folder, Virginia Department of Health, Regulatory Case Files, 1975–1997, Library of Virginia, Richmond.

91. Douglas to Kenley, August 1, 1980, box 8, folder 17, Bateman Papers.

92. Memo, James Kenley, August 26, 1980, box 10, "Regulations Prohibiting Taking of Crabs and Finfish from the Chesapeake/James River, 1980" folder, Virginia Department of Health, Regulatory Case Files, 1975–1997, LVA.

93. Proctor, *Cancer Wars*, 72.

94. "Fishing Now Legal: State Eliminates Kepone Ban," *Winchester (Va.) Star*, November 22, 1980, 10; Glenn Frankel, "Va. Removes Most of Kepone Ban on Fishing in James," *Washington Post*, November 22, 1980; and Shelley Rolfe, "Fishing Games," *Richmond Times-Dispatch*, November 22, 1980.

95. Kenley to David Rall, February 9, 1981, box 10, "Regulations Prohibiting the Taking of Crabs and Finfish from Chesapeake/James River" folder, Virginia Department of Health, Regulatory Case Files, 1975–1997, Library of Virginia, Richmond.

96. Moore to Kenley, February 27, 1981, in box 10, "Regulations Prohibiting the Taking of Crabs and Finfish from Chesapeake/James River" folder, Virginia Department of Health, Regulatory Case Files, 1975–1997, Library of Virginia, Richmond.

97. Yuzo Hayashi, "Overview of Genotoxic Carcinogens and Non-Genotoxic Carcinogens," *Experimental and Toxic Pathology* 44, no. 8 (December 1992): 465.

98. Memo, William Dykstra, EPA Office of Pesticides and Toxic Substances, August 1980, in box 10, folder: "Regulations Prohibiting the Taking of Crabs and Finfish from Chesapeake/James River," Virginia Department of Health, Regulatory Case Files, 1975–1997, Library of Virginia, Richmond; and A. E. Sirica et al., "Evaluation of Chlordecone in a Two-Stage Model

of Hepatocarcinogenesis: A Significant Sex Difference in the Hepatocellular Carcinoma Incidence," *Carcinogenesis* 10, no. 6 (June 1989): 1047–54.

99. Carolyn Click, "Virginia Health Board Lifts Kepone Ban," *Washington Post*, May 10, 1988.

100. Michael A. Unger, interview with the author, September 8, 2020.

101. Michael A. Unger, "Kepone in the James River Estuary: Past, Current and Future Trends," W&M ScholarWorks, 2017, 3, https://doi.org/10.21220/v5zw35.

102. William Goldfarb, "Changes in the Clean Water Act since Kepone: Would They Have Made a Difference?," *University of Richmond Law Review* 29, no. 3 (Spring 1995): 603–33; Library of Congress, Environmental and Natural Resources Policy Division, *A Legislative History of the Clean Water Act of 1977: A Continuation of the Legislative History of the Federal Water Pollution Control Act : Together with a Section-by-Section Index* (Washington, D.C.: U.S. Government Printing Office, 1978), http://hdl.handle.net/2027/msu.31293011625252.

103. Bocking, *Nature's Experts*, 55.

104. Kerry H. Whiteside and Robert Gottlieb, *Precautionary Politics: Principle and Practice in Confronting Environmental Risk* (Cambridge: MIT Press, 2006), 34.

105. Bocking, *Nature's Experts*, 51.

Chapter 7. The Present and Future of Kepone

1. Rob Nixon, *Slow Violence and the Environmentalism of the Poor* (Cambridge: Harvard, 2011).

2. Cedric Pietralunga, "In Martinique, Macron Calls Chlordecone Pollution an Environmental Scandal," *Le Monde*, September 27, 2018. Where news reports and other documents were offered in English, those were used. Otherwise, translations by the author.

3. Rokhaya Diallo, "A Court Case Is Shining a Light on Injustices in France's Former Colonies," *Washington Post*, February 10, 2021.

4. "Rapport: Commission d'enquête sur l'impact économique, sanitaire et environnemental de l'utilisation du chlordécone et du paraquat comme insecticides agricoles dans les territoires de Guadeloupe et de Martinique," Assemblée nationale, 2019, https://www.assemblee-nationale.fr/dyn/15/rapports/cechlordec/l15b2440-ti_rapport-enquete.

5. "Rapport, 2019."

6. "Rapport, 2019."

7. Luc Multinger, Phillippe Kadhel, and Florence Rouget, "Chlordecone Exposure and Adverse Effects in French West Indies Populations," *Environmental Science and Pollution Research* 23, no. 1 (2016): 3–8.

8. "Impacts de l'utilisation de la chlordécone et des pesticides aux Antilles: Bilan et perspectives d'évolution," Assemblée nationale, Sénat, https://www.senat.fr/rap/r08-487/r08-487.html.

9. "Rapport, 2019."

10. "Impacts de l'utilisation de la chlordécone"; and Jacques A. Bertrand et al., "Chlordecone in the Marine Environment around the French West Indies: From Measurement to Pollution Management Decisions," ICES-CM 2010 / F07, 2, Archimer, https://archimer.ifremer.fr/doc/00014/12511/9361.pdf.

11. Jean-Yves le Deaut and Catherine Procaccia, *Pesticide Use in the Antilles: Current Situation and Perspectives for Change* (Paris, France: Parliamentary Office for Scientific and Technological Assessment, 2009), https://www.senat.fr/opecst/resume/4pagesbanane_anglais.pdf.

12. "Rapport, 2019."

13. Pierre-Benoit Joly, *La saga du chlordécone aux Antilles françaises: Reconstruction chronologique 1968–2008* (Paris: Institut de Recherche sur les Fruits et Agrumes, 2010), 17–23, https://www.anses.fr/fr/system/files/SHS2010etInracol01Ra.pdf.

14. Malcolm Ferdinand, "Bridging the Divide to Face the Plantationocene: The Chlordecone Contamination and the 2009 Social Events in Martinique and Guadeloupe," in *Struggle of Non-Sovereign Caribbean Territories: Neoliberalism Since the French Antillean Uprisings of 2009* (New Brunswick, N.J.: Rutgers University Press, 2021), 67.

15. Soraya Boudia et al., "Residues: Rethinking Chemical Environments," *Engaging Science, Technology, and Society* 4 (2018): 169, https://doi.org/10.17351/ests2018.245.

16. Joly, *La saga du chlordécone aux Antilles françaises*; and "Rapport."

17. "Rapport, 2019."

18. "Rapport d'information sur l'utilisation du chlordécone et des autres pesticides dans l'agriculture Martiniquaise et Guadeloupéenne," Assemblée nationale, 2005, https://www.assemblee-nationale.fr/12/rap-info/i2430.asp.

19. Discussed in Vanessa Agard-Jones, "Bodies in the System," *Small Axe* 17, no. 3 (2013): 189.

20. Agard-Jones, "Bodies in the System," 189.

21. Le Deaut and Procaccia, *Pesticide Use in the Antilles*; and "Rapport," 2019.

22. "Le plan chlordécone IV (2021–2027)," Ministère des Solidarités et de las Santée, March 5, 2021, https://solidarites-sante.gouv.fr/sante-et-environnement/les-plans-nationaux-sante-environnement/article/le-plan-chlordecone-iv-2021-2027.

23. Magalie Lesueur Jannoyer et al., eds., *Crisis Management of Chronic Pollution: Contaminated Soil and Human Health* (Boca Raton, Fla.: CRC Press, 2016), 18.

24. Jannoyer et al., *Crisis Management*, 80.

25. Gilles Bocquene, "Pesticide Contamination of the Coastline of Martinique," *Marine Pollution Bulletin* 51, no. 5–7 (2005): 612–19; and Maynard Nichols, "Sedimentologic Fate and Cycling of Kepone in an Estuarine System: Example from the James River," *Science of the Total Environment* 97–98 (1990): 407–40.

26. Jannoyer et al., *Crisis Management*, 83.

27. Jannoyer et al., *Crisis Management*, 100, 110.

28. Jannoyer et al., *Crisis Management*, 145.

29. Philippe Kadhel, Christine Monfort, and Nathalie Costet, "Chlordecone Exposure, Length of Gestation, and Risk of Preterm Birth," *American Journal of Epidemiology* 179, no. 5 (2014): 541.

30. Marion Chevallier et al., "Natural Chlordecone Degradation Revealed by Numerous Transformation Products Characterized in Key French West Indies Environmental Compartments," *Environmental Science and Technology* 53, no. 11 (2019): 6133–43, https://doi.org/10.1021/acs.est.8b06305.

31. David Malquist, "International Research Project Reflects Mixed News on Kepone," William & Mary News, August 13, 2019, https://www.wm.edu/news/stories/2019/research-project-reflects-mixed-news-on-kepone.php.

32. Le Deaut and Procaccia, *Pesticide Use in the Antilles*.

33. Ferdinand, "Bridging the Divide to Face the Plantationocene," 58.

34. Tim Whewell, "The Caribbean Islands Poisoned by a Carcinogenic Pesticide," BBC News, November 20, 2020, https://www.bbc.com/news/stories-54992051.

35. Ferdinand, "Bridging the Divide to Face the Plantationocene," 35.

36. Ferdinand, "Bridging the Divide to Face the Plantationocene," 61.

37. Céline Tabou, "Un pesticide détruit la pêche martinquaise," *Témoignages*, December 26, 2012, https://www.temoignages.re/international/outre-mer/un-pesticide-detruit-la-peche-martiniquaise,61990.

38. Ferdinand, "Bridging the Divide to Face the Plantationocene," 63, 60.

39. Eric Stimpfling and Yasmina Yacou, "Procès chlordécone: La possibilité d'un non-lieu provoque la colère des plaignants," France TV, January 21, 2021 (accessed March 19, 2021, https://la1ere.francetvinfo.fr/guadeloupe/proces-chlordecone-la-possibilite-d-un-non-lieu-provoque-la-colere-des-plaignants-914557.html.

40. "Thousands Protest in France's Martinique against Insecticide 'Impunity,'" France 24, February 28, 2021, accessed March 27, 2021, https://www.france24.com/en/france/20210228-thousands-protest-in-france-s-martinique-against-insecticide-impunity.

41. "Pétition: Non a la prescription du crime chlordécone!," MesOpinions, https://www.mesopinions.com/petition/nature-environnement/prescription-crime-chlordecone/124508 (accessed March 19, 2021).

42. Guy Etienne, "Chlordécone aux Antilles: Ni non-lieu ni prescription... Les réactions s'enchainent," France TV, January 31, 2021, accessed March 21, 2021, https://la1ere.francetvinfo.fr/martinique/chlordecone-aux-antilles-ni-non-lieu-ni-prescription-les-reactions-s-enchainent-921523.html.

43. "Chlordécone: Quelles sont les 52 propositions du 'Lyannaj pou Dépolyé Matinik,'" January 28, 2020, accessed March 27, 2021, https://www.martinique.franceantilles.fr/actualite/environnement/chlordecone-quelles-sont-les-52-propositions-du-lyannaj-pou-depolye-matinik-551889.php.

44. Norah MacKendrick and Kate Cairns, "The Polluted Child and Maternal Responsibility in the U.S. Environmental Health Movement," *Signs: Journal of Women in Culture and Society* 44, no. 2 (2019): 307–32.

45. Jannoyer et al., *Crisis Management of Chronic Pollution*, 178.

46. Kadhel, Monfort, and Costet, "Chlordecone Exposure, Length of Gestation, and Risk of Preterm Birth"; and Margaux Maudouit and Michael Rochoy, "Revue systématique de l'impact du chlordécone sur la santé humaine aux Antilles françaises," *Therapie* 74, no. 6 (2019): 611–25, https://doi.org/10.1016/j.therap.2019.01.010.

47. Multinger, Kadhel, and Rouget, "Chlordecone Exposure and Adverse Effects in French West Indies Populations"; and Jannoyer et al., *Crisis Management*, 186.

48. "Prostate Cancer Statistics," World Cancer Research Fund, August 22, 2018, https://www.wcrf.org/dietandcancer/cancer-trends/prostate-cancer-statistics.

49. Luc Multinger et al., "Chlordecone Exposure and Risk of Prostate Cancer," *Journal of Clinical Oncology* 28, no. 21 (2010): 3457–62.

50. "Rapport, 2019."

51. Soraya Boudia et al., "Residues: Rethinking Chemical Environments," *Engaging Science, Technology, and Society* 4 (2018): 168, https://doi.org/10.17351/ests2018.245.

52. Mark Haley, interview with the author, November 4, 2016.

53. EPA *Actions against the Hopewell, Virginia, Wastewater Treatment Facility* (Washington, D.C.: U.S. General Accounting Office, 1981), 16.

54. Clint Strong, interview with the author, November 11, 2016.

55. EPA *Actions against the Hopewell, Virginia, Wastewater Treatment Facility*, 18.

56. Thomas Grubisich, "Hopewell Plant Is Probed for Pollution Violations," *Washington Post*, October 17, 1978.

57. EPA *Actions against the Hopewell, Virginia, Wastewater Treatment Facility*, 3.

58. Haley, interview.

59. Joe Patoux, "Council Accepts $25,000 Fine," *Hopewell News*, June 2, 1981.

60. Strong, interview.

61. Haley, interview.

62. Joseph McClain, "James River Population of Bald Eagles Continues to Grow, but More Slowly," *William & Mary News*, April 1, 2020.

63. Strong, interview.

64. Strong, interview.

65. Valerie Huekse, "HCIP Pulls City, Industry Together to Answer Residents' Questions," *Hopewell News*, February 25, 1993.

66. Strong, interview.

67. Huekse, "HCIP Pulls City, Industry Together," *Hopewell News*, February 25, 1993.

68. "Hopewell's Environmental Actions Praised by the EPA," *Hopewell News*, editorial, February 26, 1993.

69. Strong, interview.

70. Brandon Carwile, "Hopewell Rebrands as Eco-Friendly," *Progress-Index* (Petersburg, Va.), October 23, 2019.

71. Jeanie LeNoir Langford, interview with the author, October 4, 2016.

72. Sarah Vogelsong, "City Joins Economic Development Authority to Spur Projects," *Progress-Index* (Petersburg, Va.), July 13, 2016.

73. *Mitigation Feasibility for the Kepone-Contaminated Hopewell/James River Areas* (Washington, D.C.: Environmental Protection Agency, Office of Water and Hazardous Materials), 1978, 12.

74. "Disposal: Virginia Maintains Kepone Owned by City," *Hopewell News*, January 27, 1978.

75. "Incinerator Approved," *Hopewell News*, February 16, 1978.

76. "Burial of Pesticide at Issue in Idaho," *New York Times*, November 14, 1976.

77. "Allied Chemical's $20-Million Ordeal with Kepone," *Fortune*, September 11, 1978, 82–91.

78. "Britain Bars Importing of Kepone in Plan to Burn It Experimentally," *New York Times*, April 5, 1978.

79. "Governor to Watch Burn with Regards to Future," *Hopewell News*, April 4, 1978.

80. Bill Hoyle, "Kepone Burning in State Dropped," *Richmond Times-Dispatch*, April 7, 1978.

81. "Allied Chemical's $20-Million Ordeal with Kepone."

82. Nicholas Brown, "Kepone Waste Shipments to Germany Suspended," *Richmond Times-Dispatch*, August 29, 1978; and Bill Hoyle, "Kepone Readied for Burial," *Richmond Times-Dispatch*, September 22, 1978.

83. James Ezell, "Secret Kepone Move Hit," *Richmond Times-Dispatch*, December 5, 1978.

84. R. E. Dorer, Bureau of Solid Waste, to Fred Hughes, Director of Public Works, Hopewell, October 30, 1975; "Compliance Inspection Report, Hopewell Water Renewal, SWP192," June 11, 2001; and "Compliance Inspection Report, Hopewell Water Renewal, SWP192," June 19, 2002, all from Virginia Department of Environmental Quality Records, Richmond. "SWP" stands for Solid Waste Permit.

85. Interview with Rev. Curtis W. Harris, VCU Libraries Digital Collections, https://digital.library.vcu.edu/islandora/object/vcu%3A6150. See also Curtis West Harris Chronology, http://www.curtiswharris.com/chronology.html; James Ezzell, "Negroes, Klan Trade Shouts ant Hopewell," *Richmond Times-Dispatch*, August 7, 1966; and "Operation Begins Today," *Hopewell News*, September 7, 1966.

86. "Drainage Work Is Approved," *Hopewell News*, April 13, 1977.

87. Fred Hughes, Hopewell City Engineer, to Kenton Chestnut, State Health Department, December 8, 1982, Virginia Department of Environmental Quality, Richmond.

88. "Compliance Inspection Report, Hopewell Water Renewal, SWP271," June 11, 2001; and "Compliance Inspection Report, Hopewell Water Renewal, SWP192," June 19, 2002, both Virginia Department of Environmental Quality.

89. "Superfund" refers to what is officially the Comprehensive Environmental Response, Compensation and Liability Act (CERCLA).

90. NUS Corporation, "RCRA Facility Investigation: Task II Report: Allied Signal, Inc. Fibers Division, Hopewell Facility," 1990, Virginia Department of Environmental Quality.

91. NUS Corporation, "RCRA Facility Investigation: Task II Report."

92. NUS Corporation, "RCRA Facility Investigation: Task I Report: Allied Signal, Inc. Fibers Division, Hopewell Facility," 1990, Virginia Department of Environmental Quality.

93. NUS Corporation, "Preliminary Assessment of Allied Chemical Co. Hopewell Plant," 1984, Virginia Department of Environmental Quality.

94. Letter, Mason McElroy, AdvanSix, to Shawn Weimer, DEQ, October 28, 2020, Virginia Department of Environmental Quality.

95. Martin V. Melosi, *Effluent America: Cities, Industry, Energy, and the Environment* (Pittsburgh: University of Pittsburgh Press, 2001), 78.

96. Steve Lerner and Phil Brown, *Sacrifice Zones: The Front Lines of Toxic Chemical Exposure in the United States* (Cambridge: MIT Press, 2012).

97. Virginia Department of Health, *Cancer in Virginia: Overview and Selected Statistics* (Richmond: Virginia Department of Health, 2018), 8; https://www.vdh.virginia.gov/content/uploads/sites/71/2019/10/VCR_AnnualReport_final.pdf.

98. "State Cancer Profiles," National Cancer Institute, https://statecancerprofiles.cancer.gov/index.html.

99. Virginia Sierra Club, *Top 25 Virginia Localities with the Highest Toxic Air Emissions* (Richmond: Virginia Chapter Sierra Club, 2017); https://www.sierraclub.org/virginia/air-toxins-reports.

100. *2018 Virginia Toxics Release Inventory Report* (Richmond: Virginia Department of Environmental Quality, 2020), https://www.deq.virginia.gov/home/showpublisheddocument/4881/637484588461430000.

101. Mark Bowles and Robert Zullo, "State and Federal Authorities Converge on Hopewell Chemical Plant as Part of Undisclosed Investigation," *Richmond Times-Dispatch*, March 13, 2018.

102. Rex Springston, "Hopewell Company Agrees to $175,000 Penalty over Pollution," *Richmond Times-Dispatch*, September 19, 2019; "Stone Container Corporation Settles Air Pollution Violations," news release, Environmental Protection Agency, August 5, 2004, https://archive.epa.gov/epapages/newsroom_archive/newsreleases/9a1dd7063a2aea81852570d60070fea0.html.

103. "Hopewell City, Virginia," Quick Facts, U.S. Census Bureau, https://www.census.gov/quickfacts/fact/table/hopewellcityvirginia,VA/AFN120212.

104. Joe Maroon, interview with the author, September 28, 2016.

105. Mark Merhige, interview with the author, October 17, 2016.

106. "State of the James, 2021," James River Association, https://thejamesriver.org/stateofthejames.

107. Noah Sachs and David Flores, *Toxic Floodwaters: The Threat of Climate-Driven Chem-*

ical Disaster in Virginia's James River Watershed (N.p.: Center for Progressive Reform, 2019), 3–4, https://cpr-assets.s3.amazonaws.com/documents/VAToxicFloodwaters.pdf.

Epilogue

1. Joe Maroon, interview with the author, September 28, 2016.
2. David Paylor, interview with the author, March 8, 2017.
3. Matthew J. Strickler, *Report to Governor Ralph S. Northam on Executive Order Number Six* (Richmond, Va.: Virginia Secretary of Natural Resources, 2019), 1; and Robert Zullo, "The Man in the Middle: DEQ Director David Paylor, a Lightning Rod for Critics, Begins Serving under Fourth Governor," *Richmond Times-Dispatch*, May 13, 2018.
4. Paylor, interview.
5. Otis Brown, interview with the author, June 24, 2018.
6. William Cummings, interview with the author, March 25, 2020.
7. John Taylor, interview with the author, October 31, 2017.
8. Rex Springston, "Signs of Life after Disaster," *Richmond Times-Dispatch*, July 2, 1995.
9. Richard Stradling, "Kepone Reshaped Life on the James," *Daily Press* (Newport News, Va.), July 23, 1995.
10. Donald (Tom) Fitzgerald, interview with the author, December 5, 2016.
11. Quoted in transcript from Bernard Crutzen, *Pour quelques bananes de plus, le scandale du chlordécone* (Camera One Television, 2019).
12. Jane Tarter, interview with the author, October 20, 2016.
13. *Environmental Quality: The Ninth Annual Report of the Council on Environmental Quality* (Washington, D.C.: The Council, 1978), 178.
14. J. Brooks Flippen, *Conservative Conservationist: Russel E. Train and the Emergence of American Environmentalism* (Baton Rouge: Louisiana State University Press, 2006), 175.
15. James Hamblin, "The Toxins That Threaten Our Brains," *Atlantic*, accessed October 29, 2020, https://www.theatlantic.com/health/archive/2014/03/the-toxins-that-threaten-our-brains/284466.
16. "Our Updated Chemical Safety Law: The Lautenberg Act," Environmental Defense Fund, accessed October 30, 2020, https://www.edf.org/health/our-updated-chemical-safety-law-lautenberg-act.
17. Environmental Protection Agency, "2020 TRI Factsheet: State—Virginia," https://enviro.epa.gov/triexplorer/tri_factsheet.factsheet_forstate?pstate=VA&pyear=2020&pParent=TRI&pDataSet=TRIQ1.
18. "Risk-Screening Environmental Indicators . . . for 2020," Environmental Protection Agency, https://www.epa.gov/rsei/rsei-results-map.
19. Rachel Carson, *Silent Spring* (New York: Houghton Mifflin Harcourt, 2002), 272.
20. Josie Garthwaite, "Superweeds, Superpests: The Legacy of Pesticides," *Green Blog*, October 5, 2012, https://green.blogs.nytimes.com/2012/10/05/the-legacy-of-pesticides-superweeds-and-superpests.
21. Gabrielle Hecht, "Residue," *Somatosphere* (blog), January 8, 2018, http://somatosphere.net/2018/residue.html.
22. Paul S. Sutter, *Let Us Now Praise Famous Gullies: Providence Canyon and the Soils of the South* (Athens: University of Georgia Press, 2015), 192.

INDEX

1493 (Mann), 38
60 Minutes (TV show), 61

accountability: Allied and, 10, 68–69, 72–85, 88–93; Arrigo and, 72, 83, 89, 92–93; biocitizenship and, 68–93; carcinogens and, 74, 88–89; fear and, 86–89; Fitzgerald and, 71, 83, 85–86, 92; Godwin and, 69; health issues and, 69–77, 80–81, 85–86, 88, 92; Hundtofte and, 72–86; ignorance and, 68–69, 72–86; Jackson and, 70; labels and, 35, 73, 152, 154, 170; legal issues and, 68, 72, 74, *83*, 90–92; Life Science and, 10, 68–85, 89, 92–93; Merhige and, 89–92; Moore and, 72–86; publicity and, 72; residues and, 68, 93; uncertainty of damage and, 86–89
activism, 149, 164–67, 175
Adams, Oscar, 107
Adams v. Life Science, 122–23
AdvanSix, 172, 178–79
African Americans, 4, 9, 51, 69–70, 175–76, 179
agnotology, 68–69
Agricultural Research Service (ARS), 35
agriculture: Allied and, 4, 26, 34, 42–43, 46–47, 75, 98, 100, 109–10, 113, 158; bananas and, 41–42 (*see also* bananas); benefits of science in, 34; biocitizenship and, 75; chemical industry and, 4, 8, 26–28, 32, 34–37, 40–47, 57, 75, 109, 134–35, 146, 180, 186; DDT and, 8, 34–36, 40, 52, 57–58, 77, 88, 104, 145–46, 149; DuPont and, 28; environmental issues and, 46–47, 57; environmental management state and, 130, 132, 134–35, 146, 156; farms, 8, 13, 27–28, 35, 38–40, 47, 130, 158, 163–65; fertilizers, 1, 27–28, 31, 35–36; Guadeloupe and, 157–61, 164–65; Hopewell and, 4, 8, 19, 26–28, 32, 44, 46–47, 75, 98, 110, 113, 180; James River and, 12, 19; legal issues and, 98, 100, 109–10, 113; Life Science and, 4, 8, 26, 46–47, 75, 98, 100, 109, 113, 132, 164; Martinique and, 157–61, 164–65; National Agricultural Chemicals Association (NACA) and, 35–36; nitrogen and, 1, 31, 170, 180; phosphorous and, 180; planned obsolescence and, 185; potatoes and, 25–26, 37–44, 163–64; Virginia Senate Agricultural Committee and, 109
AgroKimicos, 159–60
air: Clean Air Act, 21, 26, 79, 96; environmental issues and, 23, 46, 51, 57–58, 61, 63; environmental management state and, 130–31; legal issues and, 94–96, 99, 110, 116; odors and, 4, 10–15, 24, 50–51, 99, 105, 116; toxic substances and, 26, 94–96, 179, 186; winds and, 14, 24, 40, 46, 84
Air Control Board, 58, 61
Air Pollution Control Board (APCB), 95–96
Aldrin, 40–43, 146, 148, 158
algae blooms, 20
Allen, James B., 53–54, 75–76, 78, 96, 106, 136, 159
Allied Chemical, 23, 24, 48, 59, 132; accountability and, 10, 68–69, 72–85, 88–93; agriculture and, 26, 34, 42–44, 158; air pollution and, 95; chlordecone and, 3, 33, 83, 88, 158–59, 162, 164; City of Hopewell settlement of, 125; corporate charter of, 4; court trial of, 114–20; crabs and, 19, 77, 122, 142, 149; *Dale Gilbert v. Allied*, 72, 91; employment demographics of, 4; environmental management state and, 131–32, 136–37, 139, 142–43, 149–50; fines against, 92, 137; history of, 30–32; Hundtofte and, 4 (*see also* Hundtofte, Virgil A.); ignorance claims of, 10, 69; indictments of, 112; insurance coverage of, 91–92; *Janice Gilbert v. Allied*, 91; *Jerry Collins v. Allied*, 72, 91; Kepone disposal and, 174–78; Kepone production and sales and, 26, 33–34, 42–44, 47–48, 51, 66, 73–80, 84–85, 96, 110–11, 158–60, 163–64; Kepone toxicity testing and, 55, 59, 61, 73–77, 90–91, 106; landfills and, 3, 5, 124, 131, 173, 175, 178; legal issues and, 23, 67, 72, 74, 89–92, 108, 110–25, 142, 150, 166; Life Science cleanup and, 108, 125, 175–78; Life Science relationship and, 3, 82–85, 114–18; nitrogen fixation and, 31; patents of, 4, 33, 46; Pebble Ammonium Nitrate Plant, 1, 3; pesticides and, 104, 186; Plastics Division, 47, 110, 113; plea of, 113; *Pruitt v. Allied*, 122; regulation and, 23, 59, 73, 83, 94, 106, 110, 131–32, 136; sewage and, 5, 97–98, 100, 112, 116; soil contamination and, 26, 158–59, 162, 177–78; tolling method and, 4, 33–34, 44, 47–48, 115, 122; toxic substances and, 5, 67, 69, 72–73, 76, 110, 131–32, 136; trial verdicts of, 114–20; Virginia settlement of, 124–25; water pollution and, 5, 10, 18, 23, 63, 110–15, 117–19, 124–25, 162, 177–78
Ameen, Jameil, 51
American Century, 8, 26; chemicals and, 33–37
American Cyanamid Company, 20
American Industrial Hygiene Association, 146
Amory, Charles, Jr., 143–44
AMSES (Medical Association for the Protection of the Environment and Health), 165–66
Anderson, R. L., 1, 3

antimony pentachloride, 3, 73
Appomattox River, 1, 19, 24, 27, 64, 172
Archer, Edmund, 27
Arrigo, Frank: biocitizenship and, 72, 83, 89, 92–93; health issues and, 72, 83, 89, 92–93, 184–85; lawsuit of, 72; legal issues and, 72, 105; Life Science and, 1, 3, 72, 83, 89, 92, 105
arsenic, 39–40, 174
Artificial Silk Company of America, 30–32, 195n14
ASSAUPAMAR (Association for the Preservation of Martinican Heritage), 166–67
Australia, 43
Austria, 43

Bagley, Richard, Sr., 144–45
Bailey, David S., 101, 104, 106, 109, 118
Bailey Creek, 3, 56, 64, 99, 101, 104–5, 176
Baker, Bernard, 90
Baker, Edward, 52
Baliles, Gerald, 130–33, 137, 154
bananas, vii; Aldrin and, 40–43, 146, 148, 158, 165; bans and, 158, 165; Bordeaux and, 43; Cavendish, 41–42; Dieldrin and, 40–41, 43, 146, 148, 158, 161; France and, 42–43; Gros Michel, 41; Guadeloupe and, 5–6, 26, 42–43, 156–61; increased demand for, 42; Martinique and, 5–6, 26, 42–43, 156–58, 161; Panama disease and, 41–42; pesticides and, 5, 26, 37, 41–44, 156–59, 161; plantations and, 6, 41, 43, 159–60, 166; root borers and, 5, 37, 40, *41*, 44; soil and, 6, 26, 157–60; United Fruit Company and, 41; water and, 5–6, 41, 157, 161; weevils and, 5, 26, 37, 40–44, 158–60, 163–64
bans: Aldrin, 146, 158; bananas and, 158, 165; cancer and, 146, 157; challenging, 65–67; crab, 122, 128, 139, 141–44, 152; Dalton and, 150–51, 174; DDT, 36; debating, 141–50; Deichmann and, 146–50; developing countries and, 160, 164; Dieldrin, 146, 158; domestic, 159; ending, 150–55; Environmental Protection Agency (EPA) and, 5, 104, 152–53; FDA and, 139; fishing, 9, 65, 67, 119, 122, 127–29, 137–56, 183; France and, 43, 161, 164; Guadeloupe and, 156, 158, 165; HCH, 43, 158; irresponsible enforcement of, 142; Martinique and, 156, 158, 165; oysters and, 9, 65, 122, 139, 141, 143; pesticides and, 5, 158, 165; United Nations and, 5; U.S. House and, 150
Barnier law, 161
Barrett Company, 31–32
Bateman, Herb: background of, 130; Deichmann and, 146–50; environmental management state and, 130, 144–48, 151, 183; Kepone bills and, 130
Battle of the Crater, 178
BBC News, 164
Beall, J. Glenn, 136
beer, 30
beetles: Colorado potato, 26, 37–40, 43–44, 134; DDT and, 39–40; France and, 39; insecticides and, 34, 37, 40, 43; Mexico and, 37; Paris green and, *38*, 39–40; pesticides and, 26, 37–40, 43–44; spread of, 37–40
Belgium, 37
Bemiss, Fitzgerald (Gerry), 20, 22
Bender, Mike, 149
benzene, 89
Bermuda Hundred, 27
Bertin, Ambroise, 164
Best Products, 121
Bhopal, India, 135
biocitizenship: agriculture and, 75; Arrigo and, 72, 83, 89, 92–93; cancer and, 71–74, 87–92; chlordecone and, *83*, 88; contamination and, 74; Environmental Protection Agency (EPA) and, 73; Guadeloupe and, 9, 167; insecticides and, 74, 76, 79, 87; Kepone and, 68–93; landfills and, 70; legal issues and, 68, 72, 74, 76, 78, *83*, 89–93; Martinique and, 9, 167; pesticides and, 68–69, 73, 76–77; politics and, 69–70, 74, 89; pollution and, 68, 93; regulation and, 73, 77, 79, 83; social aspects of, 69–72; unions and, 70; water and, 68, 80, 82, 90, 93
blood contamination, 45, 52, 55, 57, 87–89, 101, 164, 168
blue crabs, 16, 141, 143
bluefish, 15, 118–19, 141, 143
Board of Health, 131, 133, 151–54, 174
boll weevils, vii
Bordeaux, 43
Boutrin, Louis, 161–62, 166
Bowen, H., 39
Bowles, Raymond, 103
Britain, 28, 37, 40, 42, 174
Brodhead, William, 136
Brown, Otis: crisis management and, 182; environmental issues and, 59, 61, 63; environmental management state and, 129, 138, 141, 148; legal issues and, 120, 125
Broyhill, Jim, 136
Bullard, Robert, 70
Bullard, Ross, 121
Bureau of Industrial Hygiene, 53, 107
Butler, David, 106–7
Byrd, Harry F., Jr., 169
Byrd, Harry F., Sr., 7, 19, 21

Cairns, Kate, 167
calcium carbonate, vii
Calliope, 159–60
Cambusy, Bertrand, 165
Cameroon, 43, 164
Camp Lee, 29–32
canals, 12–13
cancer: bans and, 146, 157; biocitizenship and, 71–74, 87–92; chlordecone in Caribbean and, 6–7, 9, 157–59, 164–65, 167–68; environmental issues and, 55–59, 64; environmental management state and, 127–30, 137–40, 145–49, 152–54; Epstein on, 74; Guadeloupe and, 5, 157, 159, 168;

[222] INDEX

Hopewell and, 5–6, 42, 55, 58–59, 71, 87, 106, 127, 157, 168, 179, 185; International Agency for Research on Cancer, 159; Kepone and, 5–6, 42, 58–59, 64, 71, 73–74, 87–89, 127, 138–40, 145–54, 159, 184; legal issues and, 106; Martinique and, 5, 157, 164–65, 168; National Cancer Institute (NCI) and, 42, 59, 74, 138, 147–49; prostate, 130, 157, 162, 164, 168; regulation and, 5–6, 55, 57–59, 106, 127–28, 137, 140, 148, 152, 157, 165; stomach, 184; toxic substances and, 5–6, 42, 57, 64, 71–74, 90, 106, 127–28, 137–38, 146–49, 153–54, 157, 165, 179; tumors and, 154; World Cancer Research Fund, 168

Cannon, Shanklin B., 54–55

carcinogens: accountability and, 74, 88–89; environmental management state and, 138, 140, 145–54; Environmental Protection Agency (EPA) and, 150; James River and, 59, 64; threshold level on, 150

Carey, Archie Vincent, 29–30, 195n19

Caribbean. See Guadeloupe; Martinique

Carole, Francis, 166–67

carp, 56

Carroll, Jon R., 95

Carson, Rachel, 36, 57–58, 130, 186

catfish, 15–16, 56, 140

Center for International Cooperation in Agricultural Research for Development (CIRAD), 160

Centers for Disease Control (CDC), 45, 52–58, 180

chemical industry: agriculture and, 4, 8, 26–28, 32, 34–37, 40–47, 57, 75, 109, 134–35, 146, 180, 186; corporate malfeasance and, 181; Council on Environmental Quality and, 185; DDT and, 8, 34–36, 40, 52, 57–58, 77, 88, 104, 145–46, 149; Environmental Defense Fund (EDF) and, 149; environmental management state and, 135–36; extent of in U.S., 58–59; fertilizers, 1, 27–28, 31, 35–36; herbicides, 34, 36; Hopewell and, 8, 25, 26, 32–37; labels and, 35, 73, 152, 154, 170; Manufacturing Chemists Association and, 135; marketing of, 42–44; National Agricultural Chemicals Association (NACA) and, 35–36; nitrogen and, 1, 31, 170, 180; pesticides and, 36 (see also pesticides); phosphorous and, 180; Southern expansion of, 33; synthetics and, 31–36, 41; Toxic Substances Control Act and, 135–37. See also specific company

Chesapeake Bay: crabs and, 16, 140–41, 149; fishing and, 15, 17, 63, 119, 123, 141, 143–44; James River and, 14–17, 56, 63, 105, 119, 123, 137, 140–42, 144, 147, 149–50, 170, 173; oysters and, 16–17, 123, 140–41, 149; pollution and, 8, 130, 140–44, 147–50, 181; recovery goal for, 170; seafood industry and, 17, 123, 137, 141, 143–44, 147–50; sediment and, 14, 140, 142, 173; toxic substances and, 8, 137, 140; Virginia Commission of Fisheries and, 17; Virginia Environmental Endowment and, 181

Chesapeake Bay Foundation, 181

Chirac, Jacques, 43–44

chlordecone: Allied and, 3, 33, 83, 88, 158–59, 162, 164; biocitizenship and, *83*, 88; breakdowns of, 163; Calliope and, 159–60; as carcinogen, 42, 59, 64, 74, 88–89, 138, 140, 145, 147, 149–54, 159 (see also cancer); dust from, 45, 50–53, 57, 70, 77–80, 84–86, 89, 95–96, 106; environmental issues and, 64; Guadeloupe and, 5–9, 157–68, 171; half-life of, 6–7, 56, 157, 160, 168–69, 172; health issues and, 167–69; Life Science and, 3, 8, 83, 159, 162, 164–68; Martinique and, 5–9, 6–7, 157–67, 171; protesting, 164–67; rediscovering, 160–64; SEPPIC and, 159–60; synthesis of, 33; tolling and, 4, 33–34, 44, 47–48, 115, 122

Chlordecone IV Plan, 162

chlorinated hydrocarbons, 40, 74, 89, 111, 148–49

chlorine, 31, 87–88, 105, 148, 160–61

chlorofluorocarbons (CFCs), 135

Chou, Yinan, 45, 51–54, 70, 72, 104

Chronicle of a Poisoning Foretold (Confiant and Boutrin), 161–62

City Point, 27–29, 95

civil rights, 51, 70

Civil War, 12, 17–18, 27–28, 38–39, 178

Claiborne, William, 46

Clean Air Act, 21, 26, 79, 96

Clean Water Act: criminal issues and, 97, 110–11; environmental management state and, 127, 134, 155; future issues and, 169; regulation and, 4, 22, 26, 97, 110–11, 127, 155, 169; Water Pollution Control Act and, 22

coal, 26, 29, 31, 33, 177–78, 182

Cold War, 8, 26, 34–36

Cole, Norman, 105

Coleopterorum Species (Germar), 40

Collins v. Allied Chemical. See Jerry Collins v. Allied Chemical

Colombia, 42

colonialism, 6, 9, 12, 40, 158, 165, 167

Commission of Outdoor Recreation, 20

compensation: legal issues and, 33, 71–72, 129, 142, 158; restitution, 9–10, 69, 72, 94, 106, 123–24, 181

Confiant, Raphaël, 161–62, 166

Conley, Weston, Jr., 142

Connor, John T., 108, 121

Consolidated Labs, 56

conspiracy, 54, 111–19

contamination, 12; biocitizenship and, 74; blood, 45, 52, 55, 57, 87–89, 101, 164, 168; Chesapeake Bay and, 173; containment of, 95, 99, 112, 127, 176, 181; environmental issues and, 45, 47, 51, 53, 59, 61, 64, 67; environmental management state and, 127–28, 131–32, 135, 138, 141–42, 145, 148–49, 152, 155; food and, 6, 36, 112, 141, 145, 148–49, 152, 155–58, 162–63; Guadeloupe and, 157–64, 187; Hopewell and, 6–10, 22–23, 33, 36, 43, 45, 47, 53, 59, 67, 94–95, 99, 108, 112, 132, 155, 157, 176, 187; Hudson River and, 173; James River

INDEX [223]

contamination (*continued*)
and, 5, 9–10, 15, 64, 94, 138, 141, 156, 160, 162, 173; legal issues and, 84–85, 99, 108, 112, 125; Life Science and, 5, 47, 67, 94–95, 108, 131–32, 159, 162, 183; Martinique and, 157–64, 187; Second Hand Kepone, 185; sewage and, 5, 45, 99; South and, 8; slow violence of, 157; soil, 6, 36, 157–63, 176; toxic substances and, 5–6, 8, 64, 67, 131–32, 135, 155–56; urine samples, 45, 52; waste and, 4–5, 18–23, 45, 97–99, 105, 112, 124–25, 131–32, 155, 157, 161–63; water, 9–10, 22–23, 36, 67, 94, 125, 131, 135, 141, 155–58, 161–62
Continental Can, 4
Corn, Morton, 96
coronavirus, 154, *166*, 183
Costa Rica, 43
Cotton Belt, 33
Council on Environmental Quality, 185
Council on the Environment, 22, 121, 129–30, 161
Cowpasture River, 14–15
crabs: Allied and, 19, 77, 122, 142, 149; bans and, 122, 128, 139, 141–44, 152; blue, 16, 141, 143; canning, 17; Chesapeake Bay and, 16, 140–41, 149; commercial value of, 16; contamination levels for, 57; environmental management state and, 128, 138–44, 149, 152; James River and, 13–16, 122, 128, 138, 140–44, 149, 152; legal issues and, 122; Life Science and, 10, 77, 122, 149; livelihood from, 9–10, 13–16; soft-shell, 17
"crime against every citizen", *91*, 119, 208n94
Crowe, Linda, 24, 185
Cummings, Bill, 90, 108–9, 112, 121, 169, 183
Cummings Award, 146
Cuyahoga River, 181–82

Dale, Thomas, 27
Dale Gilbert v. Allied, 72, 91
Dalkon Shield Claimants Trust, 90
Dalton, John, 141, 150–51, 169, 174–75
dams, 12–13
Daniel, Pete, 35
Daniel, Robert, 136, 169
Daniels, Dominic, 71, 76
Dan River, 182
DDT, vii, 8, 57; accountability and, 88; banning of, 36; Carson and, 36; Deichmann and, 146; Douglas and, 213n65; Environmental Defense Fund (EDF) and, 149; environmental management state and, 145–46, 149; Geigy and, 146; health issues and, 34–36; Kepone and, 52; LD rating of, 77; legal issues and, 104; potato beetle and, *38*, 39–40; public information on, 58; success of, 34–36
Deichmann, William, 146–50
DeMaria, John, 154
Democrats, 7, 21–22, 71, 75–76, 130, 135–37, 167
Dent, John, 71
Denton, Janice, 30–31

Department Directorate of Sanitary and Social Affairs (DDASS), 161
Department of Environmental Quality, 23, 175–76, 179–82
Despirol, 43
DeWitt, James B., 42–43
Dieldrin, 40–41, 43, 146, 148, 158, 161
Division of Ecological Studies (DES), 56
Dize, L. Cooper, 17
Douglas, Cathleen, 121
Douglas, James, 144, 151–52, 213n65
Douglas, William O., 121
Dow-Corning, 146
Dreisbach, 55
Driver, L. R., 30
DuBose, Orben R., 96
DuPont: France and, 28–29, 160; guncotton and, 28–31; Hopewell and, 18, 28–30; James River and, 18, 28; rayon and, 31; SEPPIC and, 159–60
Dykes, Jan, 55, 65, 72
Dykes, Thurman, 51, 55, 65, 82, 84
Dykstra, William, 154

Earth Day, 22
Eckhardt, Bob, 136
ecological issues, 58; Division of Ecological Studies (DES) and, 56; environmental management state and, 144; Environmental Protection Agency (EPA) and, 171; Guadeloupe and, 158–59, 166–67; James River and, 9, 12–16, 21, 56, 101, 173; legal issues and, 101, 121; Martinique and, 158–59, 166–67; modernity, 8; pesticides and, 36 (*see also* pesticides); regulation and, 158 (*see also* regulation); Sutter on, 8, 187
Ecology Urbaine, 166
Economic Committee, 170–71
Economic Development Committee, 170
economic issues: American Century and, 8, 26, 33–37; bans and, 183 (*see also* bans); compensation, 33, 71–72, 129, 142, 158; crabs and, 16; criminal issues and, 122, 125; environmental issues and, 26, 144, 156, 162, 164, 171, 179; environmental management state and, 129, 137, 140–44, 147–50, 156; Environmental Protection Agency (EPA) and, 147; explosives and, 28, 30, 32; fishing and, 13 (*see also* fishing); French Committee on Economic Affairs, Environment and Territory, 161–62; funding, 20–23, 34, 71, 88, 108, 120–22, 125, 148–49, 155, 162, 168, 170, 174, 176, 182; Godwin and, 21; Great Depression, 31–32; guncotton and, 28–31; hurricanes and, 159; insecticides and, 186; James River and, 12–14, 21, 122, 137, 142–44, 149–50, 156, 172–73; Kepone bills and, 124, 129–35; New Deal, 32; politics and, 21, 41, 108, 129, 137, 144, 156, 164, 186; poverty, 34, 71, 143, 179; present/future issues and, 162, 164, 179–73, 179; residues and, 173–74; restaurants, 13, 88, 123, 145; restitution, 9–10, 69,

72, 94, 106, 123–24, 181; Samuelson and, 4; seafood and, 14, 120, 122, 125, 137, 142, 144, 149–50, 156; slavery and, 9, 12–13, 27–28, 40–41, 70, 158, 167; Superfund, 176; tolling and, 4, 33–34, 44, 47–48, 115, 122; toxic chemicals and, 12, 14, 26, 137, 144, 156, 179, 186; unemployment, 4, 31, 120, 141, 165; Virginia Environmental Endowment, 22, 120–22, 125, 172, 181

Ecuador, 42–43

Edmond-Mariette, Valy, 164, 167

eels, 15–16, 139–41, 152, 154, 162

Egerton, J., 38

Ehrlich, Paul, 36

Elizabeth River, 144

Emergency Planning and Community Right-to-Know Act, 135

Emmerich, Fred J., 34–35

endocrine disruptors: chlordecone/Kepone as, 165, 168

Environmental Defense Fund (EDF), 149

environmental issues: agriculture and, 46–47, 57; air, 23, 46, 51, 57–58, 61, 63; cancer and, 55–59, 64; Carson on, 36, 57–58, 130, 186; chemical industry and, 46; Chesapeake Bay and, 17, 56, 105, 119, 130, 140–44, 150, 170, 181; Clean Water Act, 4, 22, 26, 97, 110–11, 127, 155, 169; contamination and, 45, 47, 51, 53, 59, 61, 64, 67; Council on the Environment, 22; Earth Day, 22; ecological issues and, 58 (*see also* ecological issues); economic issues and, 26, 144, 156, 162, 164, 171, 179; fishing and, 64–69; France and, 6–7, 11, 47, 163; Godwin and, 58–64, 67; Hundtofte and, 46–49, 53, 58, 61, 66–67; ignorance and, 58, 64–65; James River and, 55–67; Kepone and, 45–67; landfills and, 45, 56; legal issues and, 48, 67; mitigation of, 2–3, 173–74; Moore and, 46–53, 58, 61, 67; oysters and, 9, 12–20, 56, 64–65, 109, 122–23, 138–43, 149; ozone, 128, 135, 145; pesticides and, 52, 55, 57–62, 66; politics and, 55, 64, 67, 178; pollution, 48, 57 (*see also* pollution); postwar movement for, 8–9, 12, 21, 26; regulation and, 46, 55–59, 62–64, 67; residues and, 46, 57, 67; seafood and, 68; sediment and, 55–56; sewage and, 45, 47, 53, 56, 61–63; shellfish and, 56–57, 63–64; soil and, 46, 51; State Water Control Board (SWCB) and, 55, 57, 61; Toxic Substances Control Act (TSCA), 127, 132–34, 155, 185; Toxic Substances Information Act (TSIA), 5, 131–35; U.S. Senate and, 21; Virginia Environmental Endowment, 22, 120–22, 125, 155, 169, 172, 179, 181; water and, 46, 49, 53, 55–57, 61–62, 64, 67; Water Pollution Control Act, 18–22, 105, 110–12

environmental management state: action levels and, 10, 57, 63, 118–19, 128, 137–55, 172–73; agriculture and, 130, 132, 134–35, 146, 156; air and, 130–31; Allied and, 131–32, 136–37, 139, 142–43, 149–50; Bateman and, 130, 144–48, 151, 183; Board of Health and, 131, 133, 151–54, 174; cancer and, 127–30, 137–40, 145–49, 152–54; carcinogens and, 138, 140, 145–54; chemical industry and, 135–36; Clean Water Act and, 127, 134, 155; contamination and, 127–28, 131–32, 135, 138, 141–42, 145, 148–49, 152, 155; Council on the Environment and, 22, 121, 129–30, 161; crabs and, 128, 138–44, 149, 152; ecological issues and, 144; economic issues and, 129, 137, 140–44, 147–50, 156; Environmental Protection Agency (EPA) and, 128, 136, 139–40, 144–55; "Fish Consumption Advisory" and, 127, 180; fishing and, 127–29, 137–41, 144, 151–56; Godwin and, 129–31, 137–41, 144–45, 148, 150–51; Guadeloupe and, 156; health issues and, 127–46, 150–56; ignorance and, 128; Jackson and, 132–33, 149; James River and, 127–28, 137–44, 147, 149–56; Kepone and, 127–56; landfills and, 131; lawsuits and, 142, 150–51; legal issues and, 132, 150; Life Science and, 131–32, 135–36, 143, 145, 149, 152; Martinique and, 156; Merhige and, 133, 137; oysters and, 138–43, 149; pesticides and, 127–28, 135–40, 145–56; politics and, 128–29, 137, 144–45, 150, 152, 156; pollution and, 130–31, 136, 142–43, 145, 151, 154–55; regulation and, 127–52, 155–56; residues and, 127, 137–38, 147–48, 155; seafood and, 137–56; sediment and, 140, 142; sewage and, 131, 151, 154; shellfish and, 128, 139, 147, 149; State Water Control Board (SWCB) and, 131, 134, 137, 142–43; Toxic Substances Control Act and, 135–37; water and, 127–28, 131, 134–44, 149–56

Environmental Protection Agency (EPA): action levels and, 57; Agricultural Research Service (ARS) and, 35; bans and, 5, 104, 152–53; Bateman and, 148; biocitizenship and, 73; carcinogens and, 150; consent decree and, 170; Consolidated Labs and, 56; creation of, 22; Deichmann and, 147–48; ecological issues and, 171; economic issues and, 147; environmental management state and, 128, 136, 139–40, 144–55; Executive Order 11574 and, 97; FDA and, 139; Feasibility Study for Kepone and, 173; funding from, 169; Health Effects Research Laboratory, 5, 55; James River and, 2, 173; Kenley and, 151; legal issues and, 5, 96–99, 104–13, 122, 170; mitigation report and, 2; National Academy of Sciences and, 185; NPDES and, 97–98, 104, 169; OSHA and, 67; politics and, 170; pollution and, 5, 22, 96–98, 109, 111, 136, 151, 155; raids by, 179; remaining work of, 179; residues and, 173–74; sewage and, 61, 174; Solid Waste Management Units and, 177–78; Strong and, 169–71

Environmental Quality Act, 23

Eppes, Francis, 27

Eppes, Richard, 27–28

Epstein, Samuel S., 74

European Union, 160, 165

Evans, David, 103, 129

Executive Order 11574, 97
explosives, 28, 30, 32

Facebook, 172, 185
famine, 34, 37
FBI, 90, 109, 179
Federal Reserve Board, 31
Ferdinand, Malcolm, 158
Ferguson, Warren, 76
fertility rates, 59, 168
fertilizers, 1, 27–28, 31, 35–36
Finklea, John, 58
Firestone, 4, 85
Firth, James Wilbur, Jr., 142
"Fish Consumption Advisory" (VDH), 127, 180
fishing: bans and, 9, 65, 67, 119, 122, 127–29, 137–56; catfish, 15–16, 56, 140; Chesapeake Bay and, 15, 17, 63, 119, 123, 141, 143–44; contamination levels for, 57; eels and, 15–16, 139–41, 152, 154, 162; environmental issues and, 64–69; environmental management state and, 127–29, 137–41, 144, 151–56; Florida and, 7, 15, 139; Guadeloupe and, 165; industrialization and, 17; James River and, 5, 9, 12–18, 55–69, 119, 122–22, 127–28, 137–41, 144, 152–54, 156, 183; legal issues and, 119, 122–23; livelihood from, 9–10, 13–16; Martinique and, 165; Maryland and, 7, 17, 42–43, 123, 136, 139, 143, 173, 175; pollution and, 5, 18, 21, 154, 165; regulation and, 5, 9–10, 12, 17, 64, 127–29, 139–40, 152, 156, 165; shad, 14–17, 19, 56, 139–40, 143, 180; shrimp, 15, 162; State Water Control Board (SWCB) and, 55; Virginia Commission of Fisheries and, 17; watermen and, 9–10, 13, 17, 67, 123, 128, 144, 152, 154, 156, 183
Fitzgerald, Tom: accountability and, 71, 83, 85–86, 92; environmental issues and, 50–51; health issues and, 5, 55, 85–86, 184; House hearings and, 71; Kepone shakes and, 50–51, 55; Life Science and, 5, 50, 55, 71, 83, 92; White and, 51, 85
Flint, John, 111
Flint, Michigan, 182–83
Flippen, Brooks, 185
Flores, David, 180
Florida, 7, 15, 139, 146
Floyd, W. W., 100
food: additives and, 148; contamination of, 6, 36, 112, 141, 145, 148–49, 152, 155–58, 162–63; famine, 34, 37; "Fish Consumption Advisory" and, 127, 180; population growth and, 19, 36; shortages of, 34, 36–37. *See also* seafood industry
Food and Drug Administration (FDA), 42, 57, 128, 139, 145, 147
Food Safety Council, 147
Ford, Gerald, 109, 137, 148
Fort Lee, 18, 32, 118, 169
Fortune magazine, 91
France: bans and, 43, 161, 164; Barnier law and, 161; beetles and, 39; chemical trade of, 37; Chirac and, 43–44; colonialism of, 6; demonstrations in, *166*; DuPont and, 28–29, 160; environmental issues and, 6–7, 11, 47, 163; Guadeloupe and, 42 (*see also* Guadeloupe); health issues and, 157–68, 171, 185–86; IFAC and, 43; INRA and, 159–60; Kepone shipments and, 43; legal issues and, 121, 160–61, 164–66; Macron and, 158; Martinique and, 42 (*see also* Martinique); Ministry of Agriculture and, 43–44; residual governance and, 155–56; Vincent de Lagarrique and, 44, 47, 159–60; water and, 29, 121, 160–63; West Indies of, 16, 44, 117, 157–68
Frank R. Lautenberg Chemical Safety for the 21st Century Act, 185–86
French Committee on Economic Affairs, Environment and Territory, 161–62
French National Assembly, 158
French Overseas Fruit Research Institute (IFAC), 43
French Plant Protection Service, 43
Friends of the Lower Appomattox River (FOLAR), 24, 172

Galda, Joseph, 106
Garnett, Henry D., 150
Garrett, Fred, 143
Gay, William, 143
Geigy, J. R., 40, 146
General Accounting Office, 136, 169–70
Germany, 31, 37, 39, 47, 146, 164, 175
Germar, Ernst Friedrich, 40
Giesen, James C., vii–viii, 44
Gilbert, Dale: *60 Minutes* and, 61; background of, 50; Chou and, 45, 51–52, 72, 104; health issues and, 45, 50–52, 61, 72–73, 81–82, 89, 91, 104, 116, 118; Jackson and, 45, 52; Kepone shakes and, 42, 45, 50–52, 61, 116, 118; legal issues and, 45, 72–73, 81–82, 89, 91, 104, 116, 118; Taylor and, 52
Gilbert, E. E., 33, 76
Gilbert, Janice, 50, 61, 72, 91
Gilbert v. Allied Chemical. *See Dale Gilbert v. Allied Chemical*
Glassock, James Samuel, 130
Godwin, Mills, Jr., 7; accountability and, 69; background of, 21–22; Brown on, 182; environmental issues and, 21, 58–64, 67; environmental management state and, 129–31, 137–41, 144–45, 148, 150–51; Hargis and, 144, 207n62; ignorance and, 58, 69; legal issues and, 95, 106, 108, 119, 121, 124–25; Massive Resistance and, 21; press release of, 62–64; racial issues and, 22
Gottlieb, Robert, 64, 156
Grant, Ulysses S., 28
Gravelly Run, 111
Great Depression, 31–32
Greece, 29
greed, 185
Green Party, 175
Green Revolution, 36
Greens Guadeloupe, 167

Guadeloupe: agriculture and, 157–61, 164–65; bananas and, 5–6, 26, 42–43, 156–61; bans and, 156, 158, 165; Barnier law and, 161; biocitizenship and, 9, 167; cancer and, 5, 157, 159, 168; chlordecone and, 5–9, 157–68, 171; contamination and, 157–64, 187; Department of Health and Social Development (DSDS), 161; ecological issues and, 158–59, 166–67; environmental management state and, 156; fishing and, 165; grim outlook of, 181; health issues and, 5–6, 9, 156–58, 161, 165–69, 181, 186–87; Hundtofte and, 44; hurricanes and, 159; ignorance and, 6; Kepone shakes and, 5; legal issues and, 6, 9, 165–67, 181; pesticides and, 5, 8–9, 26, 42–43, 156–63, 167, 186; politics and, 158, 164–67; pollution and, 5, 26, 157, 161–63, 167; pregnancy issues in, 168; regulation and, 160–62, 165, 168; residual governance and, 155–56; root borers and, 5, 37, 40, *41*, 44; Serva and, 167; sewage and, 131, 151, 154; soil and, 6, 8, 26, 157–63, 168; unions and, 167; water and, 5–9, 157, 161–63, 168
guncotton, 28–31
Guzelian, Philip, 88, 91

Haley, Mark, 25, 169–70
Hand Book of Explosives for Farmers, Planters (DuPont), 28
Handbook of Poisoning (Dreisbach), 55
"Happy in Hopewell" campaign, 172
Hardware Dealers' Magazine, 28
Hargis, William, 109, 143–44, 207n62
Harris, Curtis, 51, 174–76
Harrisonburg (Va.) Daily Record, 65
Harvard, 74
Havens, William, 99
Hazelwood, Thomas, 143
health issues: accountability for Kepone/chlordecone and, 69–77, 80–81, 85–86, 90–92; 164–68; African Americans and, 70; Arrigo and, 72, 83, 89, 92–93, 184–85; biocitizenship and,69–72; 164–67; blood, 45, 52, 55, 57, 87–89, 101, 164, 168; Board of Health and, 131, 133, 151–54, 174; cancer, 5 (*see also* cancer); carcinogens, 42, 59, 64, 74, 88–89, 138, 140, 145, 147, 149–54, 159; CDC and, 45, 52–58; Chou and, 45, 51–54, 70, 72, 104; contamination and, 5 (*see also* contamination); DDT and, 34–36; environmental management state and, 127–46, 150–56; fear and, 86–89; Fitzgerald and, 5, 55, 85–86, 184; Flint, Michigan and, 182–83; France and, 157–68, 171, 185–86; Gilbert and, 45, 50–52, 61, 72–73, 81–82, 89, 91, 104, 116, 118; Guadeloupe and, 5–6, 9, 156–58, 161, 165–67, 168, 181, 186–87; Hopewell and, 10, 24, 26, 66, 70, 172 (*see also* Hopewell); Jackson and, 61 (*see also* Jackson, Robert); James River and, 5–6, 19–20, 63–65, 137–41, 149–52, 154–55, 160, 163, 172, 179–80; Kenley and, 48, 52–53, 58, *60*, 61, 95, 137–41, 144, 149, 151–53; Kepone shakes, 118 (*see also* Kepone shakes); labels and, 35, 73, 152, 154, 170; LD rating and, 76–77, 104; legal issues and, 95–96, 102, 104, 106–7, 119; Love Canal and, 34, 70, 147; malaria, vii, 29, 34, 40; Martinique and, 5–6, 9, 26, 156–58, 161, 165–68, 181, 186–87; Medical College of Virginia, 42, 52–53, 55, 61, 74, 86–89, 92, 108; National Agricultural Chemicals Association (NACA) and, 35–36; OSHA and, 53, 67, 76, 96, 106–7, 131–32; pregnancy, 9, 78, 90, 158, 167–68, 184, 186; present/future issues and, 157–67, 170–78; regulation and, 5, 12, 26, 35, 64, 73, 106, 128, 131, 152, 157, 181, 186; sterility, 5, 55, 61, 65, 71, 73, 87, 89, 92, 168, 185; tobacco, 12, 22, 27, 50, 69, 128, 146, 152; uncertainty of damage and, 86–89; urine samples, 45, 52; Virginia Department of Health (VDH) and, 45, 53, 127, 132–34, 137–40, 161, 176, 178; warnings and, 9, 72–73, 116–17, 127, 149, 152–54; water, 5–6 (*see also* water)
Henry, Patrick L., 117
herbicides, 34, 36
Hercules Powder Company, 4, 24–25, 30, 32, 179
Herrington, Norman, 74
hexachlorocyclohexane (HCH), 43, 158, 161
hexachlorocyclopentadiene (HCP), 1, 3, 53, 73, 85, 105, 107, 116
Highland, Joseph, 149
Hog Island Game Refuge, 15–16
Holton, Linwood, 7, 21–22, 105, 109, 121, 130
Honeywell, 172, 178
Hooker Chemical: accountability and, 68, 72–73; Allied and, 4, 33–34, 47–48, 68, 72–73, 104, 116; environmental issues and, 47–48
Hopewell: agriculture and, 8, 26–28, 32, 44; air pollution and, 45, 51, 94–96, 110, 179, 186; cancer and, 6, 178–79; 185; as "Chemical Capital of the South", 1, 23–26, 44, 88, 171–72; chemical industry and, 8, 25, 26, 32–37; City indictments of, 112–13; conditions after Kepone, 169–80; demographics of, 1; downtown renewal of, 172; DuPont and, 18, 28–30; early history of, 27–32; Economic Committee of, 170–71; explosives and, 28, 30, 32; guncotton and, 28–31; Kepone waste disposal and, 172–78; Kepone waste treatment and, 22–23, 45, 97–106, 110–12, 116–17, 124–25, 169–70; landfills and, 3, 5, 45, 56, 70, 105, 173, 175–76, 178, 187; lawsuits and, 23, 68, 110–13, 116–17, 120, 122, 125; Local Emergency Planning Committee and, 171; map of, 2; pesticide production and, 4, 7, 11, 26, 33, 37, 55, 95, 167, 178; politics and, 26, 31–34, 41, 44; residues and, 7, 13, 18, 25–26, 32, 42, 67–68, 94, 137, 155, 157, 161, 168–69, 172–81, 186–87; soil and, 27–28, 51, 56; water pollution and, 6, 20, 25, 56, 64, 97–106, 110–12, 116–17, 124–25, 172, 175–79, 186–87
Hopewell News, 32, 45, 51, 88, 94–95, 101, 171
Hopewell Sewage Treatment Plant (HSTP), 53, 98–100, 107
Hopewell Water Renewal, 170

Houff, Sidney, 55, 71, 87
Howard, H. D., 53, 107
Hudson River, 135–36, 173
Huggett, Bob, 140
Hughes, Fred, 45, 102
Hundtofte, Virgil A.: accountability and, 72–86; background of, 46–47; corporate charter of, 4; environmental issues and, 46–49, 53, 58, 61, 66–67; Guadeloupe and, 44; Health Department and, 53, 77, 95, 102, 104; indictments of, 112; Kepone production and, 4, 47–48, 53, 75, 78–79, 82–83, 85, 94–98, 105, 110–11, 116; Kepone shakes and, 83, 116; legal issues and, 95–107, 110–18, 122, 124; Martinique and, 44; Moore and, 4, 44, 46–49, 53, 58, 61, 67, 72–86, 95, 98–105, 110, 112–18, 122, 124; as plant manager, 4; plea of, 113; Rather interview and, 61
hunting, 12
Hurricane Allen, 159
Hurricane David, 159
hydrocarbons, 40, 74, 89, 111, 148–49
hydrology, 9, 14

ignorance: accountability and, 68–69, 72–86; agnotology and, 68–69; Allied claims of, 10; environmental issues and, 58, 64–65; environmental management state and, 128; Godwin and, 58, 69; Guadeloupe and, 6; labels and, 35, 73, 152, 154, 170; legal issues and, 101, 103; Life Science claims of, 10; Martinique and, 6; public, 58; publicity and, 42, 57, 72, 122, 145, 151, 175; regulation and, 5–6, 129
industrial hygiene, 46, 53, 76, 107, 146
insecticides: beetles and, 34, 37, 40, 43; biocitizenship and, 74, 76, 79, 87; common narratives on, vii; DDT, 8, 34–36, 40, 52, 57–58, 77, 88, 104, 145–46, 149; economic issues and, 186; future issues and, 161; hydrocarbon, 74
Insley, Roy, 152
Institute Pasteur de Lille, 161
International Agency for Research on Cancer, 159
Ivory Coast, 164

Jackson, Robert: accountability and, 70; background of, 45; environmental management state and, 132–33, 149; Gilbert and, 45, 52; Kepone shakes and, 45, 52–55, 58, 60, 61, 65; legal issues and, 107, 110; Life Science and, 45, 52–55, 58, 65, 70, 107, 132, 149
Jackson River, 14–15
Jaeger, Rudolph, 74
Jamaica, 43
James River: agriculture and, 12, 19; before Kepone, 12–23; canals and, 12; carcinogens and, 59, 64; closing, 55–68, 183; colonial era and, 12; crabs and, 13–16, 122, 128, 138, 140–44, 149, 152; dams and, 12; DuPont and, 18, 28; ecological issues and, 9, 12–16, 21, 56, 101, 173; economic issues and, 12–14, 21, 122, 137, 142–44, 149–50, 156, 172–73; environmental issues and, 55–67; environmental management state and, 127–28, 137–44, 147, 149–56; Environmental Protection Agency (EPA) and, 2, 173; "Fish Consumption Advisory" and, 127; fishing and, 5, 9, 12–18, 64–69, 119, 122–22, 127–28, 137–41, 144, 152–54, 156, 183; health issues and, 5–6, 19–20, 63–65, 137–41, 149–52, 154–55, 160, 163, 172, 179–80; hydrology of, 9, 14; lawsuits and, 10, 122–23, 142, 150–51, 170; Kepone and, 5, 56, 64–65, 101, 104–5, 127–28, 137–40, 151, 155–56, 162–63, 173; oysters and, 14–16 (see also oysters); pesticides and, 5, 13, 26, 42, 44, 61–62, 69, 94–95, 127–28, 136, 138, 140, 150, 156–57, 163, 177; politics and, 19, 21; pollution and, 18–20, 23, 143, 178–80; regulation and, 5, 12, 64, 94, 126, 128, 139, 142–43, 152, 157; residues in, 13, 18, 26, 57, 67, 94, 127, 137–38, 157, 160, 172–77, 180; seafood industry and, 14, 17, 68, 109, 122–23, 137, 139, 141–44, 147–56, 183; sediment and, 5, 13–14, 26, 55–56, 64, 125, 140, 142, 160, 162, 172–73, 177, 180; sewage and, 13, 18, 20; shellfish and, 56, 63–64, 122–23, 147, 149; waste and, 97, 122, 170; watershed of, 14
James River Association (JRA), 20, 179
James River Basin, A Progress Report (League of Women Voters), 20
James River Basin, The: Past, Present, and Future (James River Project Committee), 19
James River Basin Interleague Committee, 20
James River Project Committee, 19
Janice Gilbert v. Allied, 72, 91
Jannoyer, Magalie Lesueur, 162
Janus, Murray, 113–19
Jensen, Eugene, 98, 102–5
Jensen, Soren, 213n65
Jerry Collins v. Allied, 72, 91
Jim Crow era, 69
John Randolph Medical Center, 45, 52
Johnson, Janice, 70, 79, 93
Johnson, Lyndon B., 36–37, 89
Joint Legislative Audit and Review Committee, 105
Jones, Clifton L., 47, 98–99, 101, 110
Jones, Vanessa Agard, 162
Justice Department, 108–9, 114

Kadhel, Phillippe, 163
Kali and Salz, 175
Kampmeier, Carlos, 36–37
Keavy, Steve, 83–84, 93, 105
Keiner, Christine, 143
Kelevan, 43
Kenley, James: environmental management state and, 137–41, 144, 149, 151–53; Jackson and, 52–53, 58, 60, 61, 149; legal issues and, 95; task force and, 48, 60, 61, 95, 137, 139, 144, 149
Kepone: action levels for, 10, 57, 63, 118–19, 128, 137–55, 172–73; Allied trial and, 114–20; bio-

[228] INDEX

citizenship and, 68–93; blood contamination, 45, 52, 55, 57, 87–89, 101, 164, 168; breakdowns of, 163; as carcinogen, 42, 59, 64, 74, 88–89, 138, 140, 145, 147, 149–54, 159; CDC and, 45, 52–58; DDT and, 52; dust from, 45, 50–53, 57, 70, 77–80, 84–86, 89, 95–96, 106; environmental issues and, 45–67; environmental management state and, 127–56; FDA approval of, 42; federal indictments over, 108–14; finding/not finding in air, 94–96; Guadeloupe and, 157–68; half-life of, 6, 56, 157, 160, 168–69, 172; Hopewell and, 24 (*see also* Hopewell); James River and, 5 (*see also* James River); LD rating and, 76–77; legal issues over, 43 (*see also* legal issues); marketing of, 42–44; Martinique and, 157–68; mitigation of, 2–3, 173–74; patent of, 4, 33, 46, 159; production of, 26, 33–34, 47–51, 53, 66, 74–80, 84–85, 94–107, 110–11, 115–16, 159–60, 163–64; restitution and, 9–10, 69, 72, 94, 106, 123–24, 181; seafood industry settlements and, 122–24; Second Hand, 185; settling with Virginia, 124–26; tolling and, 4, 33–34, 44, 47–48, 115, 122; toxicity of, 9, 43, 61, 72, 76–79, 84, 96, 101, 106; U.S. House hearings on, 59, 68, 71–77, 81, 85–87, 96, 105, 115, 130; U.S. Senate hearings on, 47, 51, 58, 75–76, 79, 86, 102, 105, 142; Virginia Environmental Endowment and, 120–22; water and, 97–107

Kepone Antidote, 88
"Kepone: A Portrait in Abuse" (WRVA), 65–66
Kepone bills, 130–33
Kepone shakes: *60 Minutes* and, 61; CDC and, 45, 52–58; Chou on, 45, 51–54, 70, 72, 104; Fitzgerald and, 50–51; Gilbert and, 42, 45, 50–52, 61, 116, 118; Guadeloupe and, 5; Hundtofte and, 83, 116; Jackson and, 45, 52–55, 58, *60*, 61, 65; Moore and, 51, 83, 116; symptoms of, 5, 50–52, 89; White and, 51
Kimbrough, Renate, 52
King, Martin Luther, Jr., 51, 176
Knapp, W. A., 76–77
Korean War, 33, 130
Ku Klux Klan, 176
Kulp, Jim, 129

labels, 35, 73, 152, 154, 170
landfills: African Americans and, 70, 175–76; Allied and, 3, 5, 124, 131, 173, 175, 178; biocitizenship and, 70; environmental issues and, 45, 56; environmental management state and, 131; Hopewell and, 45, 56, 70, 99, 105, 124, 173, 175–76, 178, 187; legal issues and, 99, 105, 124; Life Science and, 3, 5, 34, 45, 56, 70, 99, 105, 124, 131, 173, 176; Love Canal and, 34, 70, 147; sewage and, 5, 45, 99, 105, 175; SMWUs, 177–78; toxic substances and, 5, 34, 131, 175; waste and, 3, 5, 45, 56, 99, 105, 124, 131, 173, 175–76, 178
Lane, Arthur, 98
Langford, Jeanie, 24, 31, 195n19

Langley Air Force Base, 18
Larson, Paul, 42, 74
Larson Toxicological Studies, 74, 90, 147
lawsuits: Allied and, 68, 72, 76, 78, 89, 93, 108, 120, 122, 125, 142, 166; biocitizenship and, 68, 72, 76, 78, 89, 93; environmental management state and, 142, 150–51; Environmental Protection Agency (EPA) and, 170; France and, 165–66; Hopewell and, 6, 10, 68, 76, 108, 112, 122, 125, 150, 170; James River and, 10, 122–23, 142, 150–51, 170; Life Science and, 68, 72, 76, 89, 93, 108, 112, 122, 125, 166; politics and, 150, 166, 184; restitution and, 9–10, 69, 72, 94, 106, 123–24, 181; seafood industry and, 6, 68, 108, 112, 120, 122–23, 151; Virginia Environmental Endowment and, 120–22
LD rating, 76–77, 104
League of Women Voters, 20
Leahy, Patrick, 75, 79, 102–3
legal issues, 183; accountability, 68, 72, 74, *83*, 90–92; *Adams v. Life Science*, 122–23; agriculture and, 98, 100, 109–10, 113; aiding and abetting, 113–14, 117; air quality, 94–96, 99, 110, 116; Allied and, 23, 43, 48, 67, 72, 74, *83*, 90–91, 94, 100, 108, 113–24, 132, 150, 166; Arrigo and, 72, 105; banana workers and, 164–65; biocitizenship and, 68, 72, 74, *83*, 90–92; cancer and, 106; certification and, 123; Clean Water Act and, 97, 110–11; Code of Virginia, 65; conspiracy, 54, 111–19; contamination and, 84–85, 99, 108, 112, 125; crabs and, 122; "crime against every citizen" and, *91*, 119, 208n94; criminal fines, 92; *Dale Gilbert v. Allied Chemical*, 72, 91; ecological issues and, 101, 121; economic issues and, 122, 125; environmental issues and, 48, 67; environmental management state and, 132, 150; Environmental Protection Agency (EPA) and, 5, 67, 96–99, 104–13, 122, 150, 170; federal indictments and, 108–14; fishing and, 119, 122–23; France and, 121, 160–61, 164–65; Gilbert and, 45, 72–73, 81–82, 89, 91, 104, 116, 118; Godwin and, 95, 106, 108, 119, 121, 124–25; Guadeloupe and, 6, 9, 157, 165–67, 181; health issues and, 95–96, 102, 104, 106–7, 119; Hundtofte and, 95–107, 110–18, 122, 124; ignorance and, 101, 103; Jackson and, 107, 110; *Janice Gilbert v. Allied*, 91; *Jerry Collins v. Allied*, 91; Justice Department and, 108–9, 114; Kepone bills and, 124, 129–35; landfills and, 99, 105, 124; Life Science and, 23, 67, 72, 74, 83, 92, 94, 103, 108, 118, 122, 124, 132, 166; Martinique and, 6, 9, 157, 165–67, 181; Merhige and, 5, 89–92, 112–25, 133, 137, 179; Moore and, 95, 98–105, 110, 112–18, 122, 124; oysters and, 109, 122–23; patents, 4, 33, 39, 46, 159; pesticides and, 94–95, 102–7, 115–16; politics and, 116, 126; pollution and, 94–105, 108–20, 124; prohibition, 30; publicity and, 122; records from, 43, 72; regulation and, 6–7, 23, 94, 103–11, 115, 117, 125–26, 132, 160, 181; residues and, 94, 125; restitution, 9–10, 69, 72, 94, 106,

legal issues (*continued*)
123–24, 181; seafood industry and, 108–14, 122–25; sediment and, 99, 125; settling with Virginia, 124–26; sewage and, 4–6, 97–102, 105–12, 116, 122, 124; shellfish and, 122–23; smoking gun and, 108–14; State Water Control Board (SWCB) and, 98, 103, 119; Supreme Court, 21, 121, 150; water and, 94, 97–100, 103–6, 109–25
Lerner, Steve, 70
Lesser Antilles, 42
Letchimy, Serge, 158
Let Us Now Praise Famous Gullies (Sutter), 187
Lewis & Clark College, 56
Lewis, Frances A., 121
Lewis, Sydney, 121
Liddy, G. Gordon, 90
Life Science Products (LSP): accountability and, 10, 68–85, 89, 92–93; *Adams v. Life Science*, 122–23; agriculture and, 4, 8, 26, 46–47, 75, 98, 100, 109, 113, 132, 164; Allied contract and, 3; Arrigo and, 1, 3, 72, 83, 89, 92, 105; chlordecone and, 3, 8, 83, 159, 163–64; contamination and, 5, 47, 67, 94–95, 108, 131–32, 159, 162, 183; crabs and, 10, 77, 122, 149; environmental management state and, 131–32, 135–36, 143, 145, 149, 152; Fitzgerald and, 5, 50–51, 55, 71, 83, 92; Hundtofte and, 46 (*see also* Hundtofte, Virgil A.); ignorance claims of, 10; indictments of, 112; Jackson and, 45, 52–55, 58, 65, 70, 107, 132, 149; Kepone production of, 3–5, 67–73, 76–77, 84–85, 94–107, 112, 115–16, 159–60, 163–64; landfills and, 3, 5, 34, 45, 56, 70, 99, 105, 124, 131, 173, 176; lawsuits and, 67–68, 72–74, 89–93, 108–09, 111, 113–18, 122, 124–25; location of, 1; pollution and, 5, 23, 26, 63, 68, 94–95, 97, 99, 103, 105, 108–9, 112, 114–19, 124, 131, 162, 165, 167; regulation and, 23, 73, 83, 94, 105–6, 131, 136; sewage and, 5, 45, 97–101, 104–6, 112, 116; shutting down, 3, 5, 68, 82, 84, 96, 101–9, 115, 117, 135; soil contamination and, 8, 26, 51, 56, 159, 162, 176–77; toxic substances and, 2, 53, 73, 85, 107, 116; waterways and, 10, 23, 94, 97, 105, 173, 176
liver tumors, 154
Love Canal, 34, 70, 147
Luce, Henry, 34
Lynnhaven River, 143

McCarthy, Jerry, 22–23, 61, 121, 129–31, 169
McCollister, John, 136
MacKendrick, Norah, 167
MacKenzie, Henry W., Jr., 121
McMurran, Lewis A., Jr., 144
Macron, Emmanuel, 158
malaria, vii, 29, 34, 40
malathion, vii
Manhattan Project, 46
Mann, Charles, 38
Manufacturing Chemists Association, 135
marinas, 13, 65, 122–23, 172, 183
Marine Resources Commission, 109, 138, 144, 150

"Market Basket, The: Food for Thought" (Deichmann), 146
Marks, Hardaway, 150–51
Maroon, Joe, 181
Marshall, Virgil H., 152
Mart, Michelle, 26, 34
Martinez, Eugenio, 90
Martinique: agriculture and, 157–61, 164–65; bananas and, 5–6, 26, 42–43, 156–58, 161; bans and, 156, 158, 165; Barnier law and, 161; biocitizenship and, 9, 167; Boutrin and, 161–62, 166; cancer and, 5, 157, 164–65, 168; chlordecone and, 5–9, 157–68, 171; Confiant and, 161–62, 166; contamination and, 157–64, 187; ecological issues and, 158–59, 166–67; environmental management state and, 156; fishing and, 165; grim outlook of, 181; health issues and, 5–6, 9, 26, 156–58, 161, 165–69, 181, 186–87; Hundtofte and, 44; hurricanes and, 159; ignorance and, 6; legal issues and, 6, 9, 157, 165–67, 181; Letchimy and, 158; Macron and, 158; pesticides and, 5, 8–9, 26, 43, 156–63, 167, 186; politics and, 158, 164–67; pollution and, 5, 26, 157, 161–67; pregnancy issues in, 168; regulation and, 160–62, 165, 168; residual governance and, 155–56; residues and, 160; root borers and, 5, 37, 40, *41*, 44; Serva and, 167; sewage and, 131, 151, 154; soil and, 6, 8, 26, 157–58, 161–63, 168; unions and, 167; Vincent de Lagarrique and, 44, 47, 159–60; water and, 5–9, 157–58, 161–63, 168
Maryland, 7, 17, 42–43, 123, 136, 139, 143, 173, 175
masculinity, 9, 69–71, 87
Massive Resistance, 21
Mathews, James T., 154
Matthiesen, G. C., 8, 47–48, 75, 100
Mauldin, Erin Stewart, vii–viii
Medical College of Virginia (MCV): accountability and, 71, 86–89, 92; Brown and 61; Houff and, 55, 71; Jackson and, 53, 55, 61; Kepone testing and, 52, 74; Larson and, 42; legal issues and, 108; Samuels and, 59
mercury, 13, 58, 136, 145
Merhige, Mark, 90, 179
Merhige, Robert J., Jr.: accountability and, 89–92; background of, 89–90; certification and, 123; City of Hopewell and, 112; "crime against every citizen" and, *91*, 119, 208n94; decision of, 117–18; environmental management state and, 133, 137; legal issues and, 5, 89–92, 112–25, 133, 137, 179; Rocket Docket and, 90
methyl isocyanate, 135
Meyer, Eugene, 31
Meyer, William R., 96
Miller, Andrew, 124, 130
mitigation, 2–3, 173–74
Mitman, Gregg, 26
MODEMAS (Movement of Democrats and Ecologists for a Sovereign Martinique), 167
Monsanto, 146

Moore, William: accountability and, 72–86; background of, 46; contradictions of, 115; Life Science corporate charter and, 4; death of, 46; environmental issues and, 46–53, 58, 61, 67; Health Department and, 53, 95, 102, 104; Hundtofte and, 4, 44, 46–49, 53, 58, 61, 67, 72–86, 95, 98–105, 110, 112–18, 122, 124; indictments of, 112; Kepone production and, 4, 47–48, 50–51, 53, 75, 78–79, 83, 85, 95, 98, 101, 105, 110, 115–16, 154; Kepone shakes and, 51, 83, 116; legal issues and, 95, 98–105, 110, 112–18, 122, 124; National Toxicology Program and, 153; pleas of, 113–14; Rather interview and, 61
Moore, Vera, 46
Morgan, Cranston, 151
Morrison, Theodore, 144, 150
Moun, 164–67
Moyer, Bill, 82, 89, 116
MSX, 19
Murphy, Michelle, 26
Murphy, Thomas, 38

National Academy of Sciences, 185
National Agricultural Chemicals Association (NACA), 35–36
National Agricultural Chemicals News and Pesticide Review, 36–37
National Aniline and Chemical Company, 31
National Cancer Institute (NCI), 42, 59, 74, 138, 147–49
National Fisheries Institute, 145, 148
National Institute for Agronomic Research (INRA), 159–60
National Institute for Occupational Health and Safety, 58
National Institute of Environmental Health Science, 153
National Pollutant Discharge Elimination System (NPDES): Hopewell and, 23, 97–98, 103–5, 110, 169
National Press Club, 136
National Toxicology Program, 153
Nease Chemical, 4, 33, 47–48
Netherlands, 37, 146
Newcomen Society of the United States, 34
New Deal, 32
Newport, Christopher, 1, 19, 24, 27, 64, 172
Newport News Shipyard, 18, 109, 130, 144, 150, 154
New York Times, 74
New Zealand, 43
nitrogen, 1, 31, 170, 180
Nixon, Richard, 97, 135
Norfolk Naval Base, 18
nylon, 32

Obama, Barack, 185–86
Occidental Petroleum, 147
Occupational Safety and Health Administration (OSHA), 53, 67, 76, 96, 106–7, 131–32

odors, 4, 10–15, 24, 50–51, 99, 105, 116
Ohio State University, 42
oil crisis, 4
"Our Common Wealth: Virginians View Their Environment" (Council on the Environment), 22
oysters: bans and, 9, 65, 122, 139, 141, 143; Chesapeake Bay and, 16–17, 123, 140–41, 149; eastern, 16; environmental issues and, 56, 64–65; environmental management state and, 138–43, 149; James River and, 9, 16, 12–20, 56, 64–65, 109, 122–23, 138–43, 149; legal issues and, 109, 122–23; livelihood from, 9, 12–13; MSX and, 19; Virginia Commission of Fisheries and, 17; Virginia Oyster Packers Association, 142–43
ozone, 128, 135, 145

PALIMA (Party for the Liberation of Martinique), 166–67
Panama disease, 41–42
Paolisso, Michael, 143
parathion, 77
Paris green, *38*, 39–40
patents, 4, 33, 39, 46, 159
Paylor, David, 55–58, 66, 181–82, 185
Pearson, James, 136
pebble plant, 1, *3*, 122, 124
Pelage, Josiane Jos, 165–66
persistent organic pollutants (POPs), 5
pesticides: Allied and, 4, 10, 26, 33–34, 42–44, 55, 61, 66, 69, 73, 76, 94, 104, 106, 115–16, 136, 150, 158–59, 173, 177, 186; American Century and, 33–37; bananas and, 5, 26, 37, 41–44, 156–59, 161; bans and, 5, 158, 165; beetles and, 26, 37–40, 43–44; biocitizenship and, 68–69, 73, 76–77; breakdowns of, 163; Carson on, 186; environmental issues and, 52, 55, 57–62, 66; environmental management state and, 127–28, 135–40, 145–56; European Union and, 160; global dependency on, 9; Guadeloupe and, 5, 8–9, 26, 42–43, 156–63, 167, 186; Hopewell production of, 4, 7, 11, 26, 33, 37, 55, 95, 167, 178; James River and, 5, 13, 26, 42, 44, 61–62, 69, 94–95, 127–28, 136, 138, 140, 150, 156–57, 163, 177; legal issues and, 94–95, 102–7, 115–16; Life Science and, 4, 55, 95, 115, 159, 167; marketing of, 42–44; Martinique and, 5, 8–9, 26, 43, 156–63, 167, 186; "Our Common Wealth" report on, 22; postwar increased use of, 8; potatoes and, 26, 37–38, 40, 43–44; regulation and, 5, 7, 26, 35, 73, 94, 106, 128, 136, 145, 148, 155, 157, 160; residues and, vii–viii, 7, 26, 42, 68, 94, 148, 155, 157, 161, 172–78, 178, 186; resistance to, 36–37, 40, 43, 173, 186; sediment and, 5, 13, 26, 140, 173, 177; soil and, 8, 26, 36, 157–59, 163, 177; waterways and, 10, 94, 157, 173; weevils and, 26, 37, 41, 43–44, 158–59; world chemical trade and, 37. *See also* specific product
"Pesticides and Public Policy" (NACA), 35–36
Peterson, Torsten, 102–3

Pfizer, 146
Philip Morris, 22–23
Philippines, 43
phosphorous, 180
Pictorial History of Hopewell, Virginia (Carey), 29
Piguet, Frank, 111, 114
plantations, 12–13, 66; banana, 6, 41, 43, 158–60, 166; Dale and, 27; Eppes and, 27–28; hurricanes and, 159
poison: carcinogens, 42, 59, 64, 74, 88–89, 138, 140, 145, 147, 149–54, 159; CDC and, 45, 52–58; circle of, 69; common narratives on, vii; insecticides, vii, 34, 37, 40, 43, 74, 76, 79, 87, 161, 186; labels and, 35, 73, 152, 154, 170; LD rating and, 76–77, 104; marketing of, 42–44; pesticides, 4 (*see also* pesticides); residues and, 7, 13, 18, 25 (*see also* residues); restitution and, 9–10, 69, 72, 94, 106, 123–24, 181; slow violence of, 157; smoking gun and, 108–14; toxic substances and, 4–5 (*see also* toxic substances)
politics: active state intervention and, 6; biocitizenship and, 69–70, 74, 89; Brown on, 182–83; Cold War and, 8, 26, 34–36; colonialism, 6, 9, 12, 40, 158, 165, 167; crisis management and, 182–83; Democrats, 7, 21–22, 71, 75, 76, 130, 136–37, 167; economic issues and, 21, 41, 108, 129, 137, 144, 156, 164, 186; environmental issues and, 55, 64, 67, 178; environmental management state and, 128–29, 137, 144–45, 150, 152, 156; Environmental Protection Agency (EPA) and, 170; Ferdinand on, 158; Guadeloupe and, 158, 164–67; Hopewell and, 26, 31–34, 41, 44; James River and, 19, 21; lawsuits and, 150, 166, 184; legal issues and, 116, 126; Martinique and, 158, 164–67; new constituencies, 21–23; New Deal, 32; partisanship, 131, 136, 182–83; racism and, 22; regulation and, 64, 126, 128, 145, 152, 183, 186; Republicans, 21–22, 109, 130, 136–37, 141, 150; social relations and, 184; Virginia Environmental Endowment and, 120–22; Watergate, 4, 90; watermen and, 9–10, 67, 128, 144, 150, 152, 156, 183. *See also* U.S. House of Representatives; U.S. Senate
Politics of Cancer (Epstein), 74
Pollard, John Garland, 30
pollution: air, 21 (*see also* air); Air Pollution Control Board (APCB), 95–96; algae blooms, 20; Allied and, 5, 10, 18, 23, 26, 48, 63, 68, 94–95, 101, 108, 110, 117–19, 124, 131, 162; before Kepone, 18–21; biocitizenship and, 68, 93; Chesapeake Bay and, 8, 130, 140–44, 147–50, 181; environmental management state and, 130–31, 136, 142–43, 145, 151, 154–55; Environmental Protection Agency (EPA) and, 5, 22, 96–98, 109, 111, 136, 151, 155; fishing and, 5, 18, 21, 154, 165; future issues and, 178–82; Guadeloupe and, 5, 26, 157, 161–63, 167; Hopewell and, 10, 18, 23, 48, 68, 94–99, 103, 105, 110–12, 117, 124, 155, 161, 180–81; industrial, 18–20; James River and, 2, 5, 10, 18–23, 26, 57, 63, 68, 94–97, 111, 136, 142–43, 151, 154, 157, 162–63, 179–80; legal issues and, 94–105, 108–20, 124; Life Science and, 5, 23, 26, 63, 68, 94–95, 97, 99, 103, 105, 108–9, 112, 114–19, 124, 131, 162, 165, 167; Martinique and, 5, 26, 157, 161–67; MSX and, 19; odors and, 24; persistent organic pollutants (POPs), 5; regulation and, 5, 20, 23, 94, 105, 110–11, 117, 131, 142–43, 145, 155, 157, 161–62, 165, 180, 186; sewage and, 5, 18, 99, 112, 116; soil, 26, 157 (*see also* soil); State Water Control Board (SWCB) and, 45, 52–58; toxic substances and, 5, 57, 68, 94–97, 131, 136, 151, 155, 161, 165, 180, 186; Water Pollution Control Act, 18–22, 105, 110–12; water, 5 (*see also* water)
polychlorinated biphenyls (PCBs), 13, 135–36, 145, 147, 155, 173, 213n65
Population Bomb, The (Ehrlich), 36
population growth, 19, 36
"Potato-Beetle Catcher", 39
potatoes, vii; agriculture and, 25–26, 37–44, 163–64; Colorado beetle, 26, 37–40, 43–44, 164; Hopewell and, 26, 37–44; pesticides and, 26, 37–38, 40, 43–44; sweet, 163–64
poverty, 34, 71, 143, 179
"Powder Town Rag" (song), 29
Powhatan chiefdom, 27
Practical Entomologist, 37–38
Prairie Farmer, 38
Precautionary Politics (Whiteside and Gottlieb), 64, 156
precautionary principle, 64, 103, 128, 156, 159, 161
pregnancy, 9, 78, 90, 158, 167–68, 184, 186
Presidential Council on Environmental Quality, 135, 185
Prince George County, 27
Proctor, Robert, 68–69, 152
Pro-Gruen (For-Green), 175
prohibition, 30
prostitution, 30
protective equipment, 53, 84, 159
Pruitt v. Allied, 122
publicity, 42, 57, 72, 122, 145, 151, 175
Puckett, Roy, 175
Puerto Rico, 43

quench tanks, 3, 124

racism, 6, 33; DuPont and, 29; employment demographics and, 4; Jim Crow era and, 69–70; politics and, 22; slavery and, 9, 12–13, 27–28, 40–41, 70, 158, 167; southern culture and, 69–70
Rappahannock River, 15
Rather, Dan, 61
Rawls, J. Lewis, Jr., 133
rayon, 31
Rechem International, 174
Reed Smith LLP, 56
Reeves, John, 100–6, 116
Refuse Act Permit Program (RAPP), 97, 110–11

[232] INDEX

regulation: action levels and, 10, 57, 63, 118–19, 128, 137–55, 172–73; Allied and, 23, 59, 73, 83, 94, 106, 110, 131–32, 136; bans, 9, 65 (*see also* bans); biocitizenship and, 73, 77, 79, 83; Board of Health and, 131, 133, 151–54, 174; cancer and, 5–6, 55, 57–59, 106, 127–28, 137, 140, 148, 152, 157, 165; Clean Water Act, 4, 22, 26, 97, 110–11, 127, 155, 169; environmental issues and, 46, 55–59, 62–64; environmental management state and, 127–52, 155–56; EPA and, 5 (*see also* Environmental Protection Agency (EPA)); failures and, 6, 105, 115, 131–32, 143, 161, 168; federal, 5, 57, 73, 111; fishing and, 5, 9–10, 12, 17, 64, 127–29, 139–40, 152, 156, 165; Guadeloupe and, 160–62, 165, 168; health issues and, 5, 26, 35, 64, 73, 106, 128, 131, 152, 157, 181, 186; Hopewell and, 5, 7, 11, 23, 26, 94, 105–6, 110, 136, 155, 157; ignorance and, 5–6, 10, 58, 103, 128–29; James River and, 5–6, 9, 12, 17, 20, 23, 62, 64, 94, 126–28, 136–37, 139, 142–43, 150, 152, 155–57, 162, 180; Kepone bills and, 124, 129–35; laxity of, vii, 1, 5–7, 53–54, 83, 106–7, 181, 184; legal issues and, 6–7, 23, 94, 103–11, 115, 117, 125–26, 132, 160, 181; Life Science and, 5, 10, 23, 55, 58, 73, 77, 79, 83, 94, 103–8, 115, 117, 128, 131–32, 135–37, 156, 162, 165; Marine Resources Commission and, 109, *138*, 144, 150; Martinique and, 160–62, 165, 168; monitoring and, 7, 11, 125, 127, 137, 155, 157, 161; pesticides and, 5, 26, 35, 73, 95–96, 106, 128, 136, 145, 148, 155–57, 160; politics and, 64, 126, 128, 145, 152, 183, 186; pollution and, 5, 20, 23, 94, 105, 110–11, 117, 131, 142–43, 145, 155, 157, 161–62, 165, 180, 186; Refuse Act Permit Program and, 97, 110–11; residual governance and, 155–56; residues and, 172–77; state, 132, 142, 150; State Water Control Board (SWCB) and, 20, 23, 55, 61, 98, 103, 119, 131, 134, 137, 142–43, 169, 176; toxic substances and, 12, 26, 64, 73, 94, 106, 110, 125–28, 131–41, 148, 155–57, 183, 186; Toxic Substances Control Act (TSCA), 5, 127, 132–37, 155, 185; Toxic Substances Information Act (TSIA), 5, 127, 129, 131–35, 155; visible emission checks and, 95; Water Pollution Control Act, 18–22, 105, 110–12; waterways and, 23, 94, 157

Republicans, 21–22, 109, 130, 136–37, 141, 150

residues: accountability and, 68, 93; concept of, 7; disposing of, 172–77; economic issues and, 173–74; environmental issues and, 46, 55, 67; environmental management state and, 127, 137–38, 147–48, 155; Environmental Protection Agency (EPA) and, 173–74; Germany and, 164; Guadeloupe and, 160; Hopewell and, 5, 7, 11, 23, 26, 94, 105–6, 110, 136, 155, 157; James River and, 13, 18, 26, 57, 67, 94, 127, 137–38, 157, 160, 172, 180; legal issues and, 94, 125; liposoluble, 165; Martinique and, 160; monitoring of, 172–77; psychological, 185; remaining work for, 178–80; sediments and, 172–77; sludge and, 172–77; SWMUs and, 176–77

Resource Conservation and Recovery Act (RCRA), 134, 176

restitution: biocitizenship and, 69, 72; legal issues and, 9–10, 69, 72, 94, 106, 123–26, 181

Richmond Times-Dispatch, 134

"right to work," 33, 143

Rivers and Harbors Act, 20, 110

Robertson, A. Willis, 21

Robertson, Thomas, 36

Rocket Docket, 90

Romania, 29

Roosevelt, Franklin D., 32

root borers, 5, 37, 40, *41*, *44*

Rowe, Maurice, 129, 175

Russell, Edmund, 34

Russell, Ernie, 142

Russia, 28–29, 39, 42

Ryan, Jim, 105, 124, 129, 131–32

Sachs, Noah, 180

Samuels, Sheldon, 59

Sand, Robert, 108, 122

Santé Environnement sans Derogations (Health and Environment without Exceptions), 164–67

Saunders, James, 53

Sawyer, James G., 111, 113

Schapiro, Mark, 69

Schiebinger, Londa, 69

School of Public Health, 74

Science Advisory Committee Panel on the World Food Supply, 37

Science magazine, 74

seafood industry: *Adams v. Life Science* and, 122–23; bans and, 141–50 (*see also* bans); Chesapeake Bay and, 17, 123, 137, 141, 143–44, 147–50; crabs, 9–10 (*see also* crabs); criminal issues and, 108–14, 120, 122–24; economic issues and, 14, 122, 125, 137, 142, 144, 149–50, 156; environmental issues and, 68; environmental management state and, 137–56; federal indictments and, 108–14; "Fish Consumption Advisory" and, 127, 180; fishing and, 9–10 (*see also* fishing); James River and, 14, 17, 68, 109, 122–23, 137, 139, 141–44, 147–56, 183; lawsuits and, 6, 68, 108, 112, 120–24, 151; oysters and, *138* (*see also* oysters); *Pruitt v. Allied* and, 122–23; settling with, 122–24; shellfish, 10, 56–57, 63–64, 122–22, 128, 139–40, 147, 149; smoking gun and, 108–14; toxic substances and, 6, 14, 137, 139, 144, 148–51, 156; Virginia Seafood Council, 142, 145, 148; watermen and, 9–10, 13, 17, 67–68, 123, 125, 128, 144, 149–56, 183

Second Hand Kepone, 185

sediment, 187; Chesapeake Bay and, 14, 140, 142, 173; environmental issues and, 55–56; environmental management state and, 140, 142; James River and, 5, 13–14, 26, 55–56, 64, 125, 140, 142, 160, 162, 172–73, 177, 180; legal issues and, 99, 125; pesticides and, 5, 13, 26, 140, 173, 177; residues and, 172–77

Sellers, Christopher, 26, 46, 68
Semet-Solvay Company, 31
SEPPIC, 159–60
Serva, Olivier, 167
sewage: Allied and, 5, 97–98, 100, 112, 116; contamination of, 5, 45, 99; drains, 5, 13, 45, 105; environmental issues and, 18, 20, 45, 47, 53, 56, 61–63; environmental management state and, 131, 151, 154; Environmental Protection Agency (EPA) and, 61, 174; Guadeloupe and, 131, 151, 154; Hopewell regional treatment plant and, 169–70; Hopewell treatment plant HSTP and, 3, 5, 53, 56, 97–104, 107, 112, 116, 124; James River and, 13, 18, 20; landfills and, 5, 45, 99, 105, 175; legal issues and, 4–6, 97–102, 105–12, 116, 122, 124; Life Science and, 5, 45, 97–101, 104–6, 112, 116; Martinique and, 131, 151, 154; pollution and, 5, 18, 99, 112, 116; sludge and, 45, 56, 99–102, 105, 124, 172–77; treatment plants, 18, 110–12; waterways and, 97–98, 105, 157, 173
shad, 14–17, 19, 56, 139–40, 143, 180
sharecroppers, vii
shellfish: contamination levels for, 149; environmental issues and, 56–57, 63–64; environmental management state and, 128, 139–40, 147, 149; James River and, 56, 63–64, 122–23, 147, 149; legal issues and, 122–23; livelihood from, 10
Shown, Nicky, 61, 84, 92–93
shrimp, 15, 162
Silent Spring (Carson), 36, 57–58, 130, 186
silica, 89, 96
Simon, Bryant, 70
slavery, 9, 12–13, 27–28, 40–41, 70, 158, 167
sludge: residues and, 172–77; sewage and, 45, 56, 99–102, 105, 124, 172–77
Smith, Joseph A., 111, 113
Smith, Robert B., III, 90–91
Snyder, Dan, 109
social media, 172, 185
soil: Allied and, 26, 158–59, 162, 177–78; bananas and, 6, 26, 157–60; contamination of, 6, 36, 157–63, 176; environmental issues and, 46, 51; erosion of, 187; Guadeloupe and, 6, 8, 26, 157–63, 168; Hopewell and, 6, 8, 26–28, 36, 52, 56, 157, 168, 172, 176, 187; Life Science and, 8, 26, 51, 56, 159, 162, 176–77; Martinique and, 6, 8, 26, 157–58, 161–63, 168; pesticides and, 8, 26, 36, 157–59, 163, 177; sediment, 5–6, 13–14, 55–56, 99, 125, 140, 142, 160, 162, 172–73, 177, 180, 187; SWMUs and, 176–77
Solid Waste Management Units (SWMUs), 177–78
Solid Waste Permit (SWP), 175–76
Solvay Process Company, 31
Soviet Union, 39, 42
Spain, 43
Spears, Ellen, 70
Special Committee on Water Resources, 20
Spiess & Sohn, 43, 47, 164
Spivey, Joseph M., III, 113

Spong, William B., Jr., 21
Stamsocott Company, 30
State Water Control Board (SWCB): environmental issues and, 55, 57, 61; environmental management state and, 131, 134, 137, 142–43; legal issues and, 98, 103, 119; monitoring well of, 176; periodic inspections and, 169; pollution and, 20; research funding and, 23; sediment samples and, 55
sterility: contamination effects and, 5, 55, 61, 65, 71, 73, 87, 89, 92, 168, 185; emotional effects of, 71, 168; Rather interview and, 61
Stockholm Convention, 5
Stone Container, 179
stormwater, 180
strikes, 32, 159, 164–65
striped bass, 14–15, 19, 56, *138*, 152, 155
Strong, Clint, 24, 169–72
Stuart, Henry Carter, 30
sturgeon, 180, 191n6
sulfuric acid, 32, 94
sulfur trioxide (SO$_3$), 1, 73, 94–96
Sun Belt, 33
Superfund, 176
Supreme Court, 21, 121, 150
Sutter, Paul, 8, 187
sweet potatoes, 163–64
Swisher, Ike, 77
Switzerland, 37–40
synthetics, 31–36, 41

TAIC, 47, 110–11, 113, 118
Tarter, Jane, 184
Tatum, William, 84–85
Taylor, Edward, 92
Taylor, John, 183; accountability and, 86–91; effects documentation and, 55; Gilbert and, 52; Jackson and, 53
TEIC, 118
THEIC, 47, 110–11, 113
Timoun Mother-Child Cohort Study, 168
Tinch, Charles, 65
Tiolito, Silvio L., 33
tobacco, 12, 22–23, 27, 50, 69, 128, 146, 152
Tolbert, Lewis, Fitzgerald (law firm), 109
tolling, 4, 33–34, 44, 47–48, 115, 122
toluene, 89
Toms River Chemical, 146
Tourbillon, Pascal, 166
Toxic Chemical Program, 149
Toxic Release Inventory (TRI), 179, 186
Toxic Risk Assessment Committee, 122
toxic substances: air and, 26, 94–96, 179, 186; Allied and, 5, 42, 72–77, 90, 104, 110–11, 173–78, 186; bans and, 160 (*see also* bans); cancer and, 5–6, 42, 57, 64, 71–74, 90, 106, 127–28, 137–38, 146–49, 153–54, 157, 165, 179; Chesapeake Bay and, 8, 137, 140; contamination and, 5–10, 12, 64, 67, 74, 94–95, 125, 131–32, 135–38, 148–49,

155–57, 187; economic issues and, 12, 14, 26, 137, 144, 156, 179, 186; Hopewell and, 6–11, 24–26, 67–68, 94–105, 110, 127, 132, 136, 155–57, 178–80, 187; James River and, 5, 14, 64, 67, 126–28, 137–40, 156, 180; Kepone bills and, 124, 129–35; landfills and, 5, 34, 131, 175; LD rating and, 76–77, 104; Life Science and, 1–5, 67–79, 82–85, 94–105, 131–32, 136, 168, 176–77; pollution and, 5, 57, 94–105, 155, 161–65, 180, 186; regulation and, 11–12, 57, 64, 103–07, 110–11, 125–28, 131–45, 148–56, 160, 183, 186; remaining work for, 178–80; residues and, 172–77 (*see also* residues); seafood and, 6, 14, 137, 139, 144, 148–51, 156; visible emission checks and, 95; warnings and, 9, 72–73, 116–17, 127, 149, 152–54; waste and, 1–5, 12–13, 97–100, 104–05, 125, 155, 157, 174–78; waterways and, 9–10, 64, 97–107, 110–11, 127, 135, 137–38, 155–57, 161–63, 180, 186

Toxic Substances Advisory Council, 132
Toxic Substances Control Act (TSCA), 5, 127, 132–37, 155, 185
Toxic Substances Information Act (TSIA), 5, 127, 129, 131–35, 155
Train, Russell, 97, 136–37
Traina, Hilda, 174–75
Treacy, Dennis, 55–58, 66, 199n43
Trowbridge, Sandy, 120, 125
Troy, Anthony, 121
Truitt, Thomas, 147–48
Tubize, 30–32, 195n19
Tunney, Gene, 135
Tunney, John V., 135

unemployment, 4, 31, 120, 141, 165
Unger, Mike, 154–55, 163
Union Carbide, 135
unions: biocitizenship and, 70; Guadeloupe and, 167; Martinique and, 167; New Deal and, 32; "right to work" and, 33, 143
United Fruit Company, 41
United Nations, 5, 163–64
United Textile Workers, 32
urine samples, 45, 52
U.S. Army Corps of Engineers, 97, 110–11, 113, 173
U.S. Coast Guard, 121
U.S. Fish and Wildlife Service, 42, 59
U.S. House of Representatives: Baliles and, 130–31, 133; bans and, 150; Bateman and, 130–31; Kepone hearings of, 59, 68, 71–77, 81, 85–87, 96, 105, 115, 130; Toxic Substances Control Act and, 135–37
U.S. Senate: environmental issues and, 21; Kepone hearings of, 72–87, 96–107, 142–43; Toxic Substances Control Act and, 135–37

Velsicol Chemical, 48, 146
Venezuela, 42–43
Vietnam War, 50, 108
Vincent de Lagarrigue, 44, 47, 159–60

vinyl chloride, 89, 136
violence, 157, 159, 167
Virginia: Board of Health and, 131, 133, 151–54, 174; Code of, 65; Democrats, 7, 21–22, 71, 75, 130, 136–37, 167; Division of Industrial Development, 33; General Assembly, 20, 23, 63, 105, 131, 134; Historic Landmarks Commission, 20; Kepone bills and, 124, 129–35; Outdoor Recreation Study Commission, 20; Republicans, 21–22, 109, 130, 136–37, 141, 150; residual governance and, 155–56; response to toxic substances by, 129–35; restitution and, 124–26; settlement with, 124–26
Virginia Commission of Fisheries, 17
Virginia Department of Environmental Quality, 56
Virginia Department of Health (VDH): bans and, 150–51; Dalton and, 150–51; environmental issues and, 45, 53; environmental management state and, 127–28, 132–34, 137–40; "Fish Consumption Advisory" and, 127, 180; present/future issues and, 176, 178
Virginia Environmental Endowment (VEE): economic issues and, 22, 120–22, 125, 155, 169, 172, 179, 181; present/future issues and, 169, 172, 179; restitution and, 181
Virginia Institute of Marine Science (VIMS), 19; environmental management state and, 109, 137, 139–44, 149; Hargis and, 143–44, 207n62; Huggett and, 141; present/future issues and, 162, 172–73
Virginia Marine Resources Commission, 109
Virginia Oyster Packers Association, 142–43
Virginia Seafood Council, 142, 145, 148
Virginia Senate Agricultural Committee, 109
Virginia Tech, 23, 56
Virginia Working Watermen's Association, 152, 154

Wald, Hakrader & Ross, 147
Walsh, Benjamin Dann, 37–38
Walton, Wayne, 24–25
War Finance Corporation, 31
Warner, John W., Jr., 169
warnings: "Fish Consumption Advisory" and, 127, 180; health issues and, 9, 72–73, 116–17, 127, 149, 152–54; Jackson on, 149; labels and, 35, 73, 152, 154, 170; LD ratings and, 76–77
Washington Post, 31
Washington Star, 74
waste: contaminated, 3, 5, 12, 22, 46, 67, 99, 112, 125, 131–32, 155, 157, 161–62, 176; Hopewell and, 18, 22, 45, 56, 97–99, 104, 110–12, 116–17, 124–25, 169–77; James River treatment of, 3, 5, 18, 56, 104, 170, 172; landfills and, 3, 5, 45, 56, 99, 105, 124, 131, 173, 175–76; liquid, 1, 99; residues and, 172–77 (*see also* residues); sludge and, 45, 56, 99–102, 105, 124, 172–77; SMWUs, 177–78; solid, 22, 45, 99, 175, 177–78; toxic substances and, 4–5, 12–13, 20, 67, 69, 97–98, 101, 104, 125, 134, 140, 155, 157, 174–75, 186

INDEX [235]

water: Allied and, 12, 18, 23, 32, 108, 110–19, 143, 162, 176–78; bananas and, 5–6, 41, 157, 161; biocitizenship and, 68, 80, 82, 90, 93; Clean Water Act, 22; contamination of, 6, 9–10, 22–23, 36, 53, 67, 94, 99, 112, 125, 127, 131, 135, 141, 152, 155–58, 161–62, 176, 184; currents of, 46; drinking, 5, 7, 21, 161–62, 172, 183; environmental issues and, 46, 49, 53, 55–57, 61–62, 64, 67; environmental management state and, 127–28, 131, 134–44, 149–56; fishing and, 5, 10–13, 21, 55–57, 64–65, 67, 109, 127–28, 137–44, 149, 162, 180 (*see also* fishing); Flint, Michigan, 182–83; France and, 29, 121, 160–63; Guadeloupe and, 5–9, 157, 161–63, 168; health issues and, 5, 10, 26, 55–57, 64, 106, 127, 137–41, 149, 161–64; Hopewell and, 25–26, 29–32, 36, 41; hydrology and, 9, 14; James River and, 10 (*see also* James River); legal issues and, 94, 97–100, 103–6, 109–25; Life Science and, 5, 23, 56, 97–106, 112, 117–18, 124, 176; Marine Resources Commission and, 109, 138, 144, 150; Martinique and, 5–9, 157–58, 161–63, 168; mercury and, 13, 58, 136, 145; pesticides and, 4–6, 10, 94, 157, 173; quality of, 98, 110, 143, 160; quench tanks and, 3, 124; regulation and, 23, 94, 157; Rivers and Harbors Act, 110; sediment and, 5–6 (*see also* sediment); sewage and, 97–98, 105, 157, 173; Special Committee on Water Resources, 20; State Water Control Board (SWCB), 20, 23 (*see also* State Water Control Board (SWCB); stormwater, 180; toxic substances and, 10, 57, 94, 98, 135, 157; Virginia Marine Resources Commission and, 109; waste, 3, 18, 21–22, 49, 99–100, 104–5, 112, 134, 160, 169

Water Control Law, 18

Watergate, 4, 90

watermen: "Fish Consumption Advisory" and, 127; fishing bans and, 9, 65, 67, 119, 122, 127–29, 137–56; politics and, 9–10, 67, 128, 144, 150, 152, 156, 183; seafood industry and, 9–10, 13, 17, 67–68, 123, 125, 128, 141–44, 149–56, 183

Water Pollution Control Act, 18–22, 105, 110–12. *See also* Clean Water Act

Water Resources Research Act, 23

Water Resources Research Center, 23

weevils: Aldrin and, 40–43, 146, 148, 158; banana, 5, 26, 37, 40–44, 158–60, 163–64; boll, vii; Bordeaux and, 43; Dieldrin and, 40–41, 43, 146, 148, 158, 161; exploding population of, 159; pesticides and, 26, 37, 41, 43–44, 158–59

Weicker, Lowell, 136

Weigel, Kit, 32, 95, 172

Weir, David, 69

WestRock, 179

Weyanoke Indians, 27

White, Delbert: contamination and, 50–52, 61, 71–72, 78, 80–82, 85–86, 96, 116, 118; Kepone shakes and, 51; Rather interview and, 61

White, Pat, 72, 86

white people: employment and, 4, 6, 9, 29, 33, 69–70, 167; lower-income, 178; mortality rates and, 179; settlers, 38

white perch, 155

Whiteside, Kerry H., 64, 156

Williams, Gerald, 111, 114

Wilson, Norwood, 122, 124

Wilson, Woodrow, 31

wind, 14, 24, 40, 46, 84

workers: black, 69–70; compensation and, 33, 71–72, 129, 142, 158; health issues and, 5 (*see also* health issues); Kepone dust on, 45, 50–53, 57, 70, 77–80, 84–86, 89, 95–96, 106; labels and, 35, 73, 152, 154, 170; masculinity and, 9, 69–71, 87; OSHA and, 53, 67, 76, 96, 106–7, 131–32; restitution for, 9–10, 69, 72, 94, 96, 123–24, 181; settling cases of, 89–93 (*see also* legal issues); slavery and, 9, 12–13, 27–28, 40–41, 70, 158, 167; unions, 32–33, 70, 143, 167; white, 4, 9, 29, 33, 69–70, 167, 178

World Cancer Research Fund, 168

World War I, 28, 31

World War II 89, 128; environmental issues and, 19–20; Hopewell and, 26, 32–35, 39–44; James River and, 12; pesticide use after, 8

WRVA, 65–66

xylene, 89

Youngblood, Joan, 142

Yowell, George, 121

Yugoslavia, 43

Zero Chlordecone, 164–67

ENVIRONMENTAL HISTORY AND THE AMERICAN SOUTH

Lynn A. Nelson, *Pharsalia: An Environmental Biography of a Southern Plantation, 1780–1880*

Jack E. Davis, *An Everglades Providence: Marjory Stoneman Douglas and the American Environmental Century*

Shepard Krech III, *Spirits of the Air: Birds and American Indians in the South*

Paul S. Sutter and Christopher J. Manganiello, eds., *Environmental History and the American South: A Reader*

Claire Strom, *Making Catfish Bait out of Government Boys: The Fight against Cattle Ticks and the Transformation of the Yeoman South*

Christine Keiner, *The Oyster Question: Scientists, Watermen, and the Maryland Chesapeake Bay since 1880*

Mark D. Hersey, *My Work Is That of Conservation: An Environmental Biography of George Washington Carver*

Kathryn Newfont, *Blue Ridge Commons: Environmental Activism and Forest History in Western North Carolina*

Albert G. Way, *Conserving Southern Longleaf: Herbert Stoddard and the Rise of Ecological Land Management*

Lisa M. Brady, *War upon the Land: Military Strategy and the Transformation of Southern Landscapes during the American Civil War*

Drew A. Swanson, *Remaking Wormsloe Plantation: The Environmental History of a Lowcountry Landscape*

Paul S. Sutter, *Let Us Now Praise Famous Gullies: Providence Canyon and the Soils of the South*

Monica R. Gisolfi, *The Takeover: Chicken Farming and the Roots of American Agribusiness*

William D. Bryan, *The Price of Permanence: Nature and Business in the New South*

Paul S. Sutter and Paul M. Pressly, eds., *Coastal Nature, Coastal Culture: Environmental Histories of the Georgia Coast*

Andrew C. Baker, *Bulldozer Revolutions: A Rural History of the Metropolitan South*

Drew A. Swanson, *Beyond the Mountains: Commodifying Appalachian Environments*

Thomas Blake Earle and D. Andrew Johnson, eds., *Atlantic Environments and the American South*

Chris Wilhelm, *From Swamp to Wetland: The Creation of Everglades National Park*

Gregory S. Wilson, *Poison Powder: The Kepone Disaster in Virginia and its Legacy*

www.ingramcontent.com/pod-product-compliance
Lightning Source LLC
Chambersburg PA
CBHW011742220426
43665CB00024B/2904